WATER
Perspectives, Issues,
Concerns

WATER
Perspectives, Issues, Concerns

Ramaswamy R. Iyer

SAGE Publications
New Delhi ■ Thousand Oaks ■ London

First published in 2003 by

Sage Publications India Pvt Ltd
B-42, Panchsheel Enclave
New Delhi 110 017

Sage Publications Inc
2455 Teller Road
Thousand Oaks, California 91320

Sage Publications Ltd
6 Bonhill Street
London EC2A 4PU

Published by Tejeshwar Singh for Sage Publications India Pvt Ltd, typeset by InoSoft Systems in 10.5 pt. AGaramond and printed at Chaman Enterprises, Delhi.

Second Printing 2004

Library of Congress Cataloging-in-Publication Data

Iyer, Ramaswamy R.
    Water—perspectives, issues, concerns/Ramaswamy R. Iyer.
        p.   cm.
    Includes bibliographical references and index.
    1. Water—supply—Government policy—India. 2. Water resources development—Government policy—India. 3. Water—Law and legislation—India. 4. Dams—India. I. Title.

HD1698.I4I94   333.91'00954—dc21   2003   2002155671

ISBN: 0–7619–9759–8 (US-Hb)   81–7829–236–X (India-Hb)

Sage Production Team: Ankush Saikia, Mathew P.J., Neeru Handa and Santosh Rawat

To

Suhasini

who saw the importance of water
long before I did

Contents

List of Maps

List of Maps

Preface

This book has grown out of papers and articles on water-related issues and themes written by the author over the past 10 years and more. The attempt to weave them together into a structured whole entailed much work by way of deletion, addition, rewriting and reorganizing. It is for the reader to judge the result. This is neither a comprehensive survey on water nor a work of scholarship or research. It is a book of exposition, analysis and discussion, which covers a number of interconnected themes and issues (but, alas, also leaves out some). The discussions aim at lucidity and readability, without sacrificing depth and rigour of argument. They are addressed primarily to the interested and concerned general reader. However, the author hopes that scholars and students in various disciplines, policy makers, administrators, engineers, NGOs, activists and media persons will also find the discussions useful from their respective points of view. He hopes further that his readers, in whatever category they fall, will join him in his efforts at exploratory thinking, even if they eventually arrive at conclusions and recommendations different from his.

The book is organized in six sections. The first section is a cluster of chapters on matters arising from India's federalism: the appropriateness and adequacy of the provisions relating to water in the Constitution, and the need, if any, for amendments; the difficulties encountered in operating the constitutional/statutory mechanism for resolving inter-State river water disputes, and the solutions recommended/adopted; the nature of the current Cauvery waters dispute, why it has become so intractable and how it can be resolved; the need for a water policy at the national (i.e., federal) level, the story of the National Water Policy 1987 and a critique of the new National Water Policy adopted in 2002; and a discussion of the idea of 'river basin planning' in a federal context.

The second section relates to perceptions, perspectives and laws: different ways of perceiving water (commodity, commons, basic right, divinity); different perspectives (riparian, federalist, legal, civil society, human rights, economic and other) on water and rights, and the possibility of bringing these within an overarching framework; legislation relating to groundwater; and the question of whether there is a need for a national water law.

The third section relates to an important and vexatious controversy of our time, namely the one relating to large dams. An account of the controversy, the arguments on both sides, and the implications, is followed by a brief chapter on large dams and trans-boundary aspects; a longish chapter on the framework of laws, policies, institutions and procedures within which large-dam projects have been undertaken in India; a critique of the Supreme Court's judgement on the Narmada Bachao Andolan's writ petition on the Sardar Sarovar Project; an analysis of the dysfunctional relationship that developed between the Government of India and the World Commission on Dams; and a personal note on the evolution over the years in the author's thinking on the subject of large dams.

The fourth section consists of two chapters on the recent trend of discussing scarcities and conflicts relating to water resources in the language of security. It draws attention to the fallacies and dangers implicit in that language.

The next section entitled 'Relations with Neighbours' consists of a long chapter on conflict-resolution as illustrated by the three Treaties between India and her neighbours (the Indus Treaty with Pakistan, the Mahakali Treaty with Nepal and the Ganges Treaty with Bangladesh), followed by a short chapter on the question of a perceived shortage of waters in the Ganges and the idea of 'augmenting' the flows of that river.

The final section is concerned with planning for the future. It sets forth some basic information and concepts; provides a diagnosis of past weaknesses, problems and failures in relation to water resource planning and management; outlines a set of objectives for the future; attempts a review and critique of some recent studies of future water requirements; examines the dilemmas that confront what goes by the name of 'water resource development'; makes a set of recommendations as to what we should do to minimize, if not eliminate, those dilemmas, and finally, discusses the proposal for the linking of the rivers.

The writing is mainly in the third person, but two chapters, namely, Chapter 4 (section I) and Chapter 16 (section III), as also Appendix II, which are personal in nature, adopt the first person.

New Delhi, December 2002.   Ramaswamy R. Iyer

Some Terms Used in this Book

The expositions and discussions in this book are largely in language accessible to the general reader, and abbreviations and acronyms are explained on the first occurrence. However, the following explanations may be found useful.

1. Measures of water flows:
   Cusec = cubic feet per second.
   Cumec = cubic metres per second.
   One cumec = 35.315 cusec.

2. Measures of Volumes:
   TMC (an old measure still in use in the southern states) = Thousand million cubic feet of water.
   One TMC = 28.32 million cubic metres = 22956.8 acre-feet.
   MAF = Million acre-feet (One acre-foot is one foot of water on an acre of land).
   One MAF = 1233.48 million cubic metres.
   One hectare-metre = 8.107 acre-feet.
   BCM = Billion cubic metres.
   Km^3 = Cubic kilometers.
   One BCM = One km^3

3. Norm for water supply:
   LPCD = Litres per capita per day.

4. Land:
   MHA = Million hectares.
   One hectare = 2.47 acres.

5. Explanations of Terms:
   (i) Aquifer: 'A water-bearing stratum of permeable rock or sand or gravel' (Webster); 'a body of permeable rock able to hold

or transmit water' (Concise Oxford, 10th edition). Loosely used, particularly in writings and discussions on groundwater, in the sense of an underground water source.

(ii) Artesian well: A perpendicular well where water rises constantly to the surface under pressure, without pumping.

(iii) Afflux bund: A low embankment for holding the 'afflux' (i.e., the waters that rise because of a barrage or other structure) and preventing it from spreading.

(iv) Barrage: A structure across a river by which the water level is raised for the purpose of diversion. The difference between a barrage and a dam is that the former does not *store* waters significantly whereas the latter does.

(v) Basin: Area drained by a river and its tributaries. The basin of a river is the total area within which whatever water falls will, except for evaporation and seepage into the ground, eventually find its way to the river.

(vi) Catchment: Area from which a river or reservoir draws its water. The terms 'catchment' and 'basin' are often used interchangeably.

(vii) Embankment: 'A raised structure to hold back water or to support a roadway'. In the context of floods in rivers, embankments are structures of earth or stone constructed to contain flood waters and prevent them from spreading, and to thus protect certain areas from inundation.

(viii) Gigantism: A preference for big projects; the fascination that gigantic projects hold for some people. The term is in common use now as an extension of the biological or medical sense in which the term occurs in dictionaries.

(ix) Make-up water: If the water supply authorities were to insist on the recycling and reuse of water by industry, a small allocation of water may have to be made periodically to make up for the water that will be lost in the process of recycling. This is referred to as 'make-up water'.

(x) Pondage area: The area in which water collects behind a structure such as a barrage.

(xi) Precipitation: Fall of rain, sleet, snow or hail.

(xii) Riparian: Household, farm, settlement, village or larger political unit (including a nation), located alongside of a river (or abutting on to a river, or through which a river flows), and therefore having claims (referred to as 'riparian rights') on the river.

(xiii) Run-off: The surface-flow that results from rainfall (apart from whatever evaporates or seeps or percolates into the ground) and eventually joins the river. The river-flow is also sometimes referred to as 'run-off'.

(xiv) Transpiration: Water vapour given off by plants.

(xv) Watershed: Literally, the ridge or line of high land separating two areas so that rainwater falling on one side of the line drains on that side and cannot pass to the other side. By extension, the term is used to refer to the area bounded by the ridge. The term is generally used to denote a small local area bounded by low ridges, but a large area bounded by high hills, including a river basin, can also be described as a watershed.

(xvi) Water-harvesting: Capturing and conserving rainwater, or retarding run-off through various structures either for the direct use of stored waters or for recharging groundwater aquifers.

6. Acronyms:

CBA	Cost-Benefit Analysis.
CGWB	Central Groundwater Board.
CGWA	Central Groundwater Authority.
CPR	Common Pool Resource.
CWC	Central Water Commission.
EIA	Environmental Impact Assessment.
ETO	Exploratory Tubewells Organization.
GSI	Geological Survey of India.
MEA	Ministry of External Affairs.
MoEF	Ministry of Environment and Forests.
MoWR	Ministry of Water Resources.
NCIWRDP	National Commission for Integrated Water Resources Development Plan.
NCRWC	National Commission to Review the Working of the Constitution.
NGO	Non-Governmental Organization; also referred to as 'voluntary agency'.
NWP	National Water Policy.

NWRC	National Water Resources Council.
PIM	Participatory Irrigation Management (refers to the transfer of the management of parts of major/medium irrigation systems to farmers' associations; also known as Irrigation Management Transfer or IMT).
TAC	Technical Advisory Committee, a popular name for the Committee (set up by the Planning Commission) that clears major/medium irrigation projects for inclusion in the five-year Plans.
WCD	World Commission on Dams.
WUA	Water Users' Association.

7. <u>Note</u>:

In this book, 'state' with a small 's' is the abstract noun, i.e., the political-science concept, and 'State' with a capital 'S' refers to a State in our federal structure.

Acknowledgements

Permissions received from diverse sources for the use of material earlier published or contributed by me in various contexts are gratefully acknowledged. A list of the papers and articles from which material has been drawn (in varying degrees, ranging from indirect/negligible to direct/extensive) is given in Appendix I at the end of the book.

Over the years, my thinking about issues relating to water has been helped and influenced by exchanges, discussions and collaborative work with friends, colleagues and associates in many different contexts and forums. They have included government officials past and present, academics, World Bank officials, NGO leaders, prominent social workers, activists and campaigners, young researchers and others. To all of them, too numerous to list in full here, my thanks. Without being invidious, I must mention some with whom I have worked more closely than with others: R. Rangachari, B.G. Verghese, A. Vaidyanathan, V.B. Eswaran, Shekhar Singh, G.N. Kathpalia, L.C. Jain, John Briscoe, S. Parasuraman, Ajaya Dixit, Dipak Gyawali, Himanshu Thakkar and Radhika Gupta. I have also benefited greatly from exchanges with the late Dr Anil Agarwal, Medha Patkar, Shripad Dharmadhikari, Vandana Shiva, Anna Hazare, Rajendra Singh, Anupam Mishra and the late Vilasrao Salunke, as also Dr M.P. Vasimalai and Prof. C.R. Shanmugam of the Dhan Foundation.

I am very grateful to Dr Ajit Mozoomdar who took the trouble of reading and commenting on the entire text, and to Usha Ramanathan who performed a similar service in relation to a part of the text. The responsibility for any errors or deficiencies that remain is entirely mine.

Finally, I must express my gratitude to Sage Publications, and in particular to Omita Goyal and Ankush Saikia, for their understanding and helpfulness.

New Delhi   Ramaswamy R. Iyer
December 2002

I

Aspects of Federalism

I

Aspects of Federalism

1

Water and the Constitution of India

Entries in the Constitution

A statement often made is that under the Indian Constitution water is a State subject. There is a tendency to take this proposition for granted as a basic datum from which to proceed to further propositions and arguments. Those further propositions and arguments take one of two directions: one is to assert that water is rightly a State subject, that this position must be accepted and that the Centre must refrain from encroaching into this area; the other is to deplore that water is a State subject and to argue that the Centre needs to play an important role in regard to this precious resource, and that in order to facilitate this water should be transferred to the Concurrent List. Both these views are over-simplifications, as will be clear from the following examination of the constitutional provisions.

The relevant provisions are Entry 17 in the State List, Entry 56 in the Union List and Article 262. There are other articles and entries which may have a bearing on the matter, but the ones just mentioned are specifically concerned with water.

Entry 17 in the State List runs as follows:

Water, that is to say, water supplies, irrigation and canals, drainage and embankmènts, water storage and water power subject to the provisions of Entry 56 of List I.

It can be seen at once that it is not an unqualified entry. Water is indeed in the State List but this is subject to the provisions of Entry 56 in the Union List, which runs as follows:

Regulation and development of inter-State rivers and river valleys to the extent to which such regulation and development under the control of the Union is declared by Parliament by law to be expedient in the public interest.

The legislative competence of the State Governments under Entry 17 of the State List remains unfettered only because Parliament has not made much use of the powers vested in it by Entry 56 of the Union List. It is, therefore, not quite right to say simply that water is a State subject; it is potentially as much a Central subject as a State subject, particularly as most of the country's important rivers are inter-State.

Moreover, we must also note the provisions of Article 262:

262. Adjudication of disputes relating to waters of inter-State rivers or river valleys.

(1) Parliament may by law provide for the adjudication of any dispute or complaint with respect to the use, distribution or control of the waters of, or in, any inter-State river or river valley.

(2) Notwithstanding anything in this Constitution, Parliament may by law provide that neither the Supreme Court nor any other court shall exercise jurisdiction in respect of any such dispute or complaint as is referred to in clause (1).

It stands to reason that the legislative competence of a State under Entry 17 must be exercised in such a manner as not to prejudice the interests of other States and create a water dispute within the meaning of Article 262. This has been clearly stated in some of the Tribunals' awards.

Role of the Centre

Water is not in the Concurrent List; but it is in both the Union List and the State List. The role given to the Centre in regard to inter-State rivers and river valleys is at least potentially an important one; and this is reinforced by the use of the provisions of Entry 20 in the Concurrent List, namely, 'economic and social planning', by virtue of which major and medium irrigation, hydropower, flood control and multipurpose projects have been subjected to the requirement of Central clearance for inclusion in the national Plan. This has been questioned by some State Governments, but the clearance requirement remains; and there is of course the requirement of Central clearances under the Forest Conservation Act and the Environment Protection Act. It could be plausibly argued that even under the present dispensation the Centre has significant responsibilities in relation to water, and that it has not in fact discharged those responsibilities adequately.

The River Boards Act 1956, passed by Parliament under Entry 56 of the Union List, provides only for the establishment of advisory boards, but no boards even of an advisory kind have been set up under the Act; indeed the Act has remained virtually inoperative. (This point will be discussed further in the chapter on river basin planning.) The Inter-State Water Disputes Act 1956, enacted under Article 262 of the Constitution, has also run into difficulties in recent years; this too will be taken up for a separate discussion later.

73rd and 74th Amendments

Apart from the Union and the States there is now a third tier in the constitutional structure, created by the 73rd and 74th Amendments, namely, local bodies of governance at the village and city level: the village *panchayats* and the city *nagarpalikas* (municipalities/corporations). The Eleventh and Twelfth Schedules to the Constitution lay down lists of subjects to be devolved to the *panchayats* and *nagarpalikas*. The lists include, *inter alia*, drinking water, water management, watershed development and sanitation. It seems likely that in future this third tier will come to play an important role in relation to water resource development. However, the processes of decentralization and devolution are still evolving, and the role of the third tier is yet to emerge fully.

Deficiencies of the Existing Position?

In light of the above account, can it be said that the present constitutional position in relation to water is satisfactory? The Sarkaria Commission on Centre-State relations thought so, but serious doubts in this regard seem warranted, though these are perhaps a matter of hindsight. First, even the most general entry regarding water, namely, Entry 17 in the State List, quickly slips into specific uses of water such as water supply, irrigation, etc. Secondly, irrigation looms large; and the reference to canals, embankments, drainage, water storage, and so on, shows the heavy influence of the engineering point of view. Thirdly, while the word 'water' may doubtless be taken to include groundwater, there is no specific reference to the latter; the Constitution-makers seem to have been thinking mainly of river waters. Fourthly, the Centre has been given a role only in relation to inter-State rivers and river valleys, but it is conceivable that even in a river which flows entirely in one State that State's intervention might produce environmental or social consequences in another State, and such interventions in intra-State surface waters may also have an impact on groundwater aquifers cutting across State boundaries. There is no explicit recognition of this in the Constitution. Fifthly, the constitutional provisions do not show any direct evidence of a perception of water as a natural resource, much less of water as a part of the larger environment or the ecological system. (Some of the emerging concerns were incorporated into the Constitution at a later stage. Under the 42nd Amendment of 1976, references to the protection of the environment, forests and wildlife were introduced *via* Articles 48A and 51A, and two entries relating to forests and wildlife were added to the Concurrent List.) There is also no explicit evidence of an awareness of traditional community-managed systems of rainwater harvesting or water management, or of the role of civil society in these matters. Nor is there any *overt* reference to water as a basic essential for life, and therefore a basic human and animal right.

Some of these perceptions and concerns are of relatively recent origin, and perhaps the makers of the Constitution cannot be faulted for not having foreseen these developments. Further, a Constitution provides a foundation for the laws of the land, and is essentially a *legal* document; it cannot be expected to spell out sectoral policies in detail. Subject to those caveats, however, it is possible to argue that if the kinds of thinking that have now come to prevail had been well established when the Constitution was being drafted, the constitutional provisions might well have been very different.

Amendments Needed?

However, that is a speculative reflection, and (as stated earlier) a case of hindsight. The reality is the text of the Constitution as it exists. Even if amendments to put 'water' in the Concurrent List are considered desirable, such amendments would be enormously difficult to put through and would go counter to the persistent trend towards greater decentralization and federalism.

Moreover, it is necessary to ask ourselves what precisely will be achieved by shifting water to the Concurrent List, assuming that this proves politically feasible. Such a change will merely mean that the Centre will be enabled to legislate on water. In seeking to bring water into the Concurrent List, the Central Ministry of Water Resources is essentially trying to enlarge its own role on the ground that this will serve some useful national purposes. That argument is not wholly without force, but its importance should not be exaggerated. In the first place, there is no ground for believing that the Centre will necessarily take a more holistic view of water than the States; at both levels, limited engineering-dominated perceptions tend to prevail. Secondly, the Constitution only deals with the legislative (and correspondingly the executive) powers of *governments* at the State and Central levels; and water is not a matter merely for governments. The growing movement for a revival of the 'dying wisdom' of traditional water-harvesting and water management systems and practices envisages an enlargement of the role of the community, and a transformation of the relationship between the state and civil society. This will not necessarily be facilitated by an alteration of the relationship between the Central and State Governments. From that point of view, the debate about shifting water to the Concurrent List may well appear to be merely a question of power-sharing among administrative structures at different levels.

Finally, in the context of the advocacy of community management of common pool resources, there arises the whole question of what has come to be known as 'legal pluralism', i.e., the relationship between the formal law of the statute books and 'customary law'. From this point of view again, constitutional amendments to bring water into the Concurrent List will be of no great significance.

It is not being argued that the debate about the constitutional entries relating to water is pointless. That question needs to be considered, not merely with reference to the narrow issue of the role of the Central Government, but with some of the larger perspectives mentioned above in view. The Report of the National Commission for Reviewing the Working of the Constitution (NCRWC) is silent on this issue. It is difficult to believe that the question did not come up before the Commission. Perhaps its silence represents a deliberate decision. If so, one can only regret it. It is not clear when another opportunity to go into this matter will present itself. Meanwhile, much can be done to promote a holistic view of water, better Centre-State and inter-State relations and a constructive relationship between state and civil society, even within the ambit of the existing constitutional entries relating to water.

National Water Resources Council: Statutory Backing?

A related point is that the National Water Resources Council (NWRC), an important element in Indian federalism in relation to water resources, is only an institution established by a Resolution of the Government of India and has no statutory backing. Its prestige and influence are derived from its composition with the Prime Minister as its Chairman, the Union Minister of Water Resources as Vice-Chairman and all State Chief Ministers and several Central Ministers as Members. The National Water Policy (NWP) 1987 (and now 2002) approved by the NWRC is not a law; it has only the force of consent. It may be added that the NWRC meets very infrequently, and apart from the approval of the NWP in 1987 and of the NWP 2002 15 years later, it cannot be said to have done much. It is sometimes suggested that the NWRC and the NWP should be given a statutory backing, but it is not clear whether this is in fact necessary, and if so how, and under what entries in the Constitution, this can be done. (See chapter 10 Section II, on the question of a possible national water law.)

2

Inter-State Water Disputes Act 1956: Difficulties and Solutions

This chapter will first set forth the difficulties experienced in the operation of the Inter-State Water Disputes Act 1956 (ISWD Act), the criticisms of the adjudication system that are usually put forward and the Sarkaria Commission's recommendations. It will then give an account of the amendments to the ISWD Act passed by Parliament early in 2002 and attempt a critique of those amendments. Finally, it will take note of the recent recommendations of the National Commission to Review the Working of the Constitution (NCRWC) on the subject. It may be added that there has not yet been any significant change in the position as seen in the earlier years; the amendments to the Act are very recent and it is too early to say what impact they will have.

The Statutory Machinery

Article 262 of the Constitution provides for Parliamentary legislation for the adjudication of inter-State river water disputes, and for barring the jurisdiction of the courts (including the Supreme Court) in such cases. In pursuance of these provisions there is a Parliamentary enact-ment, namely, the ISWD Act. These are important features of Indian federalism and provide an essential mechanism for conflict-resolution in relation to river waters.

Experiences in Recent Years

Initially these provisions seemed to be working well: the Krishna, Godavari and Narmada Tribunals' Awards can be regarded as successful instances of the operation of this conflict-resolution machinery. The fact that new differences have arisen, or that public interest litigation arose in some cases, does not invalidate that statement. However, the machinery no longer seems to be working satisfactorily. In the Ravi–Beas case, political difficulties in implementing the award led to a further reference being made to the Tribunal (as provided for in the ISWD Act), and 15 years after the award was given, the matter is still before the Tribunal. In the case of the Cauvery dispute, adjudication has been running a troubled course. An Interim Order given by the Tribunal in 1991 generated a secondary dispute which was resolved in a not wholly satisfactory manner, but the main dispute still remains before the Tribunal. Twelve years after the Tribunal was set up the end of the process is not yet in sight; nor do we know whether the Tribunal's final award, when it is given, will mark the end of the dispute or will give rise to further trouble.

Disenchantment

All this has led to a measure of disenchantment with the constitutional and statutory provisions. The criticisms are broadly of two types, procedural and fundamental, as encapsulated below:

(i) There are enormous delays at every stage, namely, the establishment of a tribunal; the proceedings of the tribunal and the giving of an award; the processes of further references and supplementary clarifications or orders; and the notification of the award.

(ii) Adjudication is an unsatisfactory way of dealing with such disputes; a negotiated settlement is infinitely superior; adjudication is divisive and leads to exaggerated claims by both sides; water disputes have a strong technical content and are not best handled by judges; in the absence of guidelines there are possibilities of arbitrariness on the part of the tribunals, and they could impose unfair or unworkable awards; and finally, there are no provisions to ensure that an award once given is duly implemented.

Fundamental Criticisms

Let us take the fundamental criticisms first. There is no doubt that mutual agreement is superior to adjudication as a means of resolving a dispute, whether about river waters or anything else; but when negotiations fail, conflicts still need to be resolved, and Article 262 and the ISWD Act provide a machinery for this. Some such machinery is surely very necessary.

It has been argued that the Indian system tends to jump straight from negotiations to adjudication, and that it is desirable to proceed through the intermediate stages of conciliation and mediation before taking recourse to adjudication. However, the existing provisions do not rule out recourse to negotiations, or to mediation or conciliation or arbitration. In fact, the Act requires the Central Government to satisfy itself that a negotiated settlement is not possible before setting up a tribunal, and the Central Government tends to overdo this; it can hardly be said to establish tribunals in great haste. The routes of negotiation, mediation and conciliation are open and can be tried either before, or even alongside of, a recourse to adjudication. (Arbitration is of course an *alternative* to adjudication.)

As for divisiveness, this is not an ineluctable concomitant of adjudication. Each State will undoubtedly argue its case as strongly as possible before the tribunal, but this can be done without acrimony. If goodwill and a desire to find a solution exist, adjudication need not be adversarial in spirit. On the other hand, if a spirit of accommodation were absent, negotiations, mediation and conciliation are not likely to succeed.

It seems undesirable to prescribe various mandatory stages such as mediation, conciliation, etc., through which the parties must pass in a sequence before taking recourse to adjudication: that would only introduce even greater delays. Instead, the processes of negotiation, conciliation and mediation could continue even after a tribunal is established (without suspending the adjudication proceedings), and if agreement is reached it could be reported to the tribunal and the tribunal could make that agreement its award. There have been such instances in the past.

The point that inter-State water disputes are not best dealt with by judges because the issues involved are not legal but technical has not much merit. Such disputes often involve questions of principles relating to water-sharing and water rights, and the consideration of these matters by tribunals is not very different from a discussion of legal issues. Besides, it is a false notion that judges deal with questions of law alone; the judiciary does indeed interpret the law but it also has another function, namely, the rendering of justice, based on a thorough, fair and objective examination of both facts and the law. Further, water disputes are not unique in having a technical content. All kinds of disputes which go to the courts—criminal cases, property disputes, accident cases, insurance claims, disputes arising from contracts for large projects, industrial disputes, questions of medical ethics and many other kinds of cases—may involve technical issues. It does not follow that all such cases should be taken out of the purview of the judges and entrusted to specialists. Moreover, in the absence of the ISWD Act and tribunals set up under the Act, disputes relating to river waters would have gone to the ordinary courts anyway!

It is wrong to assume that in the absence of 'guidelines' laid down by the Central Government, or Parliament, tribunals have been arbitrary. They have generally tried to take all the available 'principles' into account, such as existing judicial pronouncements national and international, international conventions such as the Helsinki Rules and so on; and each tribunal takes note of the observations of earlier tribunals; and thus a body of case law gets built up. It seems doubtful whether an attempt to frame a fresh set of guidelines will really advance matters much. If such an exercise were in fact undertaken, it seems very possible that it might take years, perhaps decades, for a national consensus to be reached; meanwhile the conflict-resolution process cannot be suspended. That statement is amply borne out by the fact that the National Water Resources Council (NWRC) has twice considered a set of draft principles but has not been able to reach any kind of consensus, or even an approach to it.

Besides, any such statement of principles on river-sharing is bound to be very broad and general, and will be of little assistance in settling particular disputes. Despite the existence of such a document, the specifics of each case and the precise manner in which the general principles will apply in the given case could still be matters for negotiation or adjudication.

As for the fear that a tribunal may impose an unfair or unworkable award, this seems groundless, and in any case there is a remedy. Within 90 days from the pronouncement of an Award (interim or final) any of the States concerned or the Central Government can make a reference back to the tribunal asking for clarifications or raising issues, and the tribunal then has to give clarifications or issue a supplementary report.

Procedural Criticisms

Turning to procedural criticisms, the basic complaint is that the processes take years to go through. There are enormous delays at every stage.

Delays in the establishment of a tribunal occur because the Central Government has first to satisfy itself that a negotiated settlement is not possible; the exploration of this possibility sometimes takes years. Delays in the finalization of the proceedings before a tribunal are in some instances the result of the continuing exploration of an agreed settlement, but even in the absence of this factor the proceedings are exceedingly slow.

Possible Time-Limits at Various Stages

The Sarkaria Commission had recommended a time-limit of one year for the Central Government to establish a tribunal upon request by a State Government, and five years for the tribunal to give its award. The recommendations were quite modest, but they remained unimplemented for several years. It was only recently (early in 2002) that certain amendments to the ISWD Act were passed by Parliament. We shall consider these amendments, but first we must take a look at the problem of non-implementation.

The Problem of Non-Implementation

The most difficult problem—and this goes beyond procedure—is that no effective sanctions are available against the contingency of non-implementation of a tribunal's award by one of the party States. The ISWD Act says that the award is 'final and binding', and provides no appeal even to the Supreme Court. It was no doubt assumed that those impressive words 'final and binding' would be truly operative; and the barring of the jurisdiction of the Courts must have been motivated by the desire to obviate protracted litigation. Perhaps the possibility of non-compliance with a tribunal's award did not occur to the makers of the Constitution. But if that were to happen, what remedies are available?

An ISWD tribunal can only give an award; it has no role to play in its implementation. Apart from the fact that once the tribunal has given its final award it will cease to exist, it has no powers of enforcement (of its interim award, if any) even when it is in existence. It has not been clothed with powers of punishment for 'contempt'. In the event of non-implementation of an ISWD tribunal's award by a State Government, the Central Government can (failing persuasion) issue a direction to the erring State and then invoke Article 356, but that seems an extreme step;

besides, when a popular government comes back it may once again refuse to implement the award. There is no easy answer to this problem.

Sarkaria Commission's Recommendation

The Sarkaria Commission had recommended that a tribunal's award should be given the status of a decree of the Supreme Court by appropriate legislation or constitutional amendment. The assumption is that no one would disobey an order of the Supreme Court. This recommendation has finally been accepted and implemented through the recent Amendment Act. We shall discuss these recommendations, but meanwhile, the author would like to place before the readers an alternative proposition put forward by him some years ago.

An Alternative Proposal

In providing for the barring of the jurisdiction of the courts the Constitution-makers doubtless wished (as mentioned earlier) to avoid the possibility of protracted litigation. However, it seems a little strange that a conflict-resolution mechanism should provide only for a single non-appealable judgement. Under this system one or more of the parties to the dispute could well be left with a sense of grievance and injustice. (It is of course true that an arbitral award is also normally non-appealable, but recourse to arbitration is a voluntary act, whereas adjudication under the ISWD Act is not, at least for one party). This writer would suggest the consideration of the possibility of allowing an appeal to the Supreme Court against the Award of a tribunal set up under the ISWD Act. (The alternative of providing for an appellate tribunal on top of the primary tribunal, as in the case of some other kinds of tribunals, will not really solve the problem; it will merely add to the delays without necessarily ensuring implementation).

The introduction of a provision for an appeal to the Supreme Court would require an amendment to the ISWD Act. Would it also require an amendment to the Constitution? In the author's view it would not; but there can be a difference of opinion on this, depending on the manner in which Article 262 is understood. Is it an enabling article or a mandatory one? It uses the word 'may' twice: Parliament *may* enact legislation for the adjudication of river water disputes; and it *may* bar the jurisdiction of the courts. In these two instances, does the word 'may' have the force of 'shall'? If not, Parliament is merely enabled (not required) to pass legislation for the establishment of tribunals, and it is also enabled (not required) to bar the jurisdiction of the courts. If this view is correct, then the provision of an appeal to the Supreme Court would require an amendment to the ISWD Act, but not to the Constitution. However, if we take the view that in both uses, or at any rate in the second one, 'may' means 'shall', and that if Parliament does pass any legislation for the establishment of tribunals it must necessarily bar the jurisdiction of the courts, then obviously the provision of an appeal to the Supreme Court will require amendments both to the ISWD Act and to the Constitution. (The NCRWC is of the view that Article 262 is only an enabling provision. This is covered in the final sub-section of this chapter.) Whatever view we may take on this matter, the requirement of an amendment to the Constitution need not prevent us from considering a proposition if it has merit.

Some may feel that it has no merit. There may be the fear that if an appeal were provided for, every case would go to the Supreme Court. Indeed it may, but that is happening even now, though on issues other than the water dispute proper. (It cannot be said that in such cases the Supreme Court has been scrupulously refraining from going into water-related issues). If an appeal to the Supreme Court is possible, at least no State can reasonably nurse a sense of grievance. As the Supreme Court's decisions are still respected and obeyed in this country, the non-implementation problem will disappear: that would be a great advantage. There need be no additional delays, as the extra time needed for an appeal to be heard and decided by the Supreme Court could be found from the time saved through the imposition of time-limits at various stages.

The above proposal did not find favour with the National Commission for Integrated Water Resources Development Plan, before which it was placed by this author as a member. The author is now placing it before a wider audience for consideration. (We shall revert to this at the end of the present chapter, after taking note of the recommendations of the NCRWC.)

Amendments to the ISWD Act 2002

As mentioned earlier, the Sarkaria Commission's recommendations, which remained unattended to for some years, were finally considered by a sub-committee as well as the Standing Committee of the Inter-State Council, and by the Inter-State Council itself. Based on the recommendations of the Inter-State Council, certain amendments have been made to the ISWD Act. The amendments prescribe certain time-limits:

- One year for the establishment of a tribunal by the Central Government on a request from a State Government;
- Three years for the tribunal to give its Award (extendable, if found necessary, by a further period not exceeding two years by the Central Government); and
- One year for the tribunal to give a further report if a reference is made to it as provided for in the ISWD Act (this one year being further extendable if necessary, with no limit specified for such extension).
- A further amendment states that the decision of the tribunal shall have the same force as an order or decree of the Supreme Court. (There are also some amendments regarding database, etc., which we need not go into here.)

It would appear that all problems experienced so far have now been tackled through these amendments, and that the future operation of this conflict-resolution machinery should be smooth. There is certainly some ground for satisfaction that the Sarkaria Commission's recommendations have at long last been implemented, and in one respect even improved upon (the five-year time limit for the tribunal to give its award proposed by the Sarkaria Commission has been reduced to three years). However, some misgivings still remain:

(i) Consider the following wording: 'When any request under Section 3 is received ... and the Central Government is of opinion that the water dispute cannot be settled by negotiations, the Central Government shall, within a period not exceeding one year from the date of receipt of such request, by notification in the Official Gazette, constitute a Water Disputes Tribunal....' Does that convey the sense that once a State Government has made a request for the establishment of a tribunal, the Central Government cannot delay establishing the same beyond a year under any circumstances, and that the process of exploring the route of negotiations must be limited to this period? Is it not possible to take the view that the establishment of a tribunal is *conditional* on the Central Government's opinion that the dispute cannot be resolved through negotiations, and that if at the end of one year the Central Government continues to be of the view that the dispute *can* be settled through negotiations, it is not obliged to establish a tribunal? Perhaps this is a misreading of the amendment, but there is certainly some ambiguity present here.

(ii) The prescription of a time-limit of three years for a tribunal to give its Award is diluted by the provision for the grant of an extension of up to two years. Perhaps some such provision was required, but why two years? Would not a provision for a one-year extension have been adequate? Be that as it may, one fears that this provision will be invoked in every case, and that in practical terms five years, and not three, will be the normal limit. One hopes that this fear will prove unwarranted. An outside limit of five years for the Award would of course be an improvement on the present situation, but would a future case in fact reach finality in five years? Consider the next point.

(iii) For a further report by the tribunal upon a reference being made to it by the Central or a State Government after it has given its award, the tribunal is given one year. Perhaps six months would have been adequate. The more important point is that this period of one year is extendable, and that there is no limit on such extensions. There is reason to fear that a further report will be asked for in every case, and that the one-year period will also be extended in every case. This is an open-ended position: all future cases could drag on exactly as past cases did. What then has been achieved by the amendments? (The Ravi–Beas Tribunal, to which a post-award reference was made in 1987, has not yet given its further report, whatever be the reasons. What is to prevent that example from being followed in future cases?)

(iv) No time-limit has been prescribed for the publication of the tribunal's decision in the Gazette by the Central Government. It might have been useful to lay down that the Central Government should notify the tribunal's decision forthwith or at any rate within, say, two weeks. In the absence of such a provision, what would happen if the Central Government were to find it politically inconvenient to notify the Award in a given case?

(v) Consider the new sub-section (2) of Section 6: 'The decision of the Tribunal, after its publication in the Official Gazette by the Central Government under sub-section (1), shall have the same force as an order or decree of the Supreme Court.' Please note that this 'force' comes into play only after the notification in the Gazette. If there are delays in the notification, the Award will have no 'force'. Leaving that aside, and noting that this amendment is based on a recommendation of the Sarkaria Commission (to deal with the non-implementation problem discussed earlier), what exactly does this mean? How can an Order not passed by the Supreme Court have the force of (i.e., virtually become) an Order of the Supreme Court? Can Parliament say so by legislation? Let us leave that to the experts and consider the intention, which is, clearly, that the tribunal's award should have the force of law and be fully and promptly obeyed. However, this is already stated in sub-section (1) which says that the decision of the tribunal 'shall be final and binding'. Non-implementation of a tribunal's decision is in fact an act of disobedience of the laws of the land. Will the decision become *more* 'final and binding' if it had the force of an Order of the Supreme Court? (Can there be different degrees of being 'final and binding'?) What if a State were to disobey or refuse to implement a tribunal's decision? Does the new sub-section mean that it can be held to be in contempt of the Supreme Court? That in fact seems to be the only additional implication that follows from the new sub-section. Will the Supreme Court entertain such a contempt petition in relation to an Order not passed by it? We shall have to see precisely how this will work. If the non-implementation problem disappears because of this amendment, that will be a matter for great satisfaction.

(vi) There is a provision that disputes already decided by tribunals before the Amendment Act shall not be reopened. It is not quite clear why such a provision was considered necessary. From one point of view, it merely states the obvious: the Amendment Act will not by itself warrant the reopening of old settled cases. On the other hand, can a decision be frozen for all time? New disputes may well arise after some years in respect of settled cases, and at some stage it may become necessary to establish a new tribunal. Should this possibility be ruled out altogether?

(vii) Finally, the point made earlier that a single, non-appealable decision may leave one party with a sense of grievance remains. The suggestion made above of a provision for an appeal to the Supreme Court still seems worth considering.

Subject to those questions and doubts, one must welcome these amendments, and hope that they will enable future cases to move faster.

Recommendations of the NCRWC

The recommendations of the NCRWC have now become available. The NCRWC is evidently not too happy about the exclusion of the jurisdiction of the Supreme Court. They have recommended the repeal of the ISWD Act and the enactment of a new Act. They feel that Article 262 is only an enabling article, and that Parliament was not obliged to enact a law for the establishment of tribunals to adjudicate inter-State river water disputes. They would like to shift river water disputes back from the tribunals to the Supreme Court. Their recommendation is to bring such disputes within the original and exclusive jurisdiction of the Supreme Court. At the same time, they are not recommending the repeal of Article 262, as it may be needed if we want to get back to the tribunals! They feel that the shift from tribunals to the Supreme Court could be brought about by fresh Parliamentary legislation after repealing the ISWD Act.

It is the author's view that the repeal of the ISWD Act would be singularly ill advised. It would be wrong to burden the Supreme Court with *original* jurisdiction on river water disputes which involve not only (or primarily) questions of law, but masses of facts and enormous technical complexities, and which are likely to drag on for years. (One is not sure whether it would be possible to prescribe time-limits to the Supreme Court.) Judicial tribunals (with technical assistance) are the right answer. (A further point is that if there are several such disputes at the same time, a tribunal can be set up in each case; it seems hardly possible to create several Benches of the Supreme Court to hear multiple disputes simultaneously.) Article 262 together with the ISWD Act represents a very useful mechanism for dispute resolution (as a last resort when negotiations fail), and it would be a great pity to dismantle it. There have been some deficiencies in the functioning of that mechanism, and the recent amendments seek to remedy them. Further improvements may be required, as pointed out in this chapter. At the same time, a single, non-appealable decision will certainly leave one or more parties to the dispute with a sense of injustice or at least with a measure of dissatisfaction. This defect in the present system will not be remedied by shifting such disputes to the exclusive, original jurisdiction of the Supreme Court, as that too leaves no room for an appeal. The NCRWC is right in wanting to restore the jurisdiction of the Supreme Court, but it should be *appellate* jurisdiction, as proposed by the author earlier in this chapter.

3

The Cauvery Dispute

The River

The Cauvery (more appropriately the Kaveri) is one of the important rivers of southern (peninsular) India. It is a system of rivers consisting of the Cauvery itself and a number of tributaries such as the Hemavati, the Kabini, the Bhavani, the Amaravati and others. Karnataka and Tamil Nadu are the principal States in the Cauvery basin, but a small part of the basin is in Kerala, and at the very end, the Cauvery delta includes Karaikal which is a part of the Union Territory of Pondicherry. Thus, the Cauvery is an 'inter-State' river in terms of the provisions of the Indian Constitution.

Brief History of the Dispute

The dispute over the sharing of Cauvery waters has a long history going back to the 19th century. The parties then were the Madras Presidency in British India and the princely State of Mysore. After prolonged discussions there was an agreement in 1892, followed by a further dispute, an arbitration under the auspices of the Government of India, an appeal against that arbitral award to the then Secretary of State in London, and, at the instance of the Secretary of State, a resumption of mutual negotiations between the two parties leading to an agreement in 1924. The details of that agreement, and the question whether it still subsists and binds the present Karnataka and Tamil Nadu (which cannot be wholly identified with the old Mysore and Madras), need not be gone into here. It is clear in any case that for the last three decades there has been a dispute between the two States which requires a fresh solution; and Kerala and Pondicherry—which were not parties to the 1924 agreement—are also involved in the present dispute.

Talks between Karnataka and Tamil Nadu went on intermittently for over two decades from the 1970s but produced no results. The Government of India made unsuccessful efforts to bring about an agreement. Based on a report (1972) by a Fact-Finding Committee appointed by the Government of India, and further studies by an Expert Committee, an agreement was worked out in August 1976. The main elements of that attempted agreement were as follows: the existing utilization of Cauvery waters was determined as 671 TMC (thousand million cubic feet), comprising 489 TMC by Tamil Nadu, 177 TMC by Karnataka, and 5 TMC by Kerala; savings were to be made in water-use by both Tamil Nadu (100 TMC) and Karnataka (25 TMC) over a period of 15 years; those savings were to be redistributed as 4 TMC to Tamil Nadu, 87 TMC to Karnataka, and 34 TMC to Kerala; and an inter-State Cauvery Valley Authority was to be established. The 'agreement' was even announced in Parliament. Unfortunately, the announcement was premature. Tamil Nadu was then temporarily under Central rule, and it was felt that the agreement should wait for a popularly elected government; and when an elected government took over, it refused to ratify the understanding because it was considered not wholly satisfactory. The prospects of a resolution of the dispute receded.

Thereafter the Central Government continued to make efforts to settle the dispute, and there were also discussions at the level of the Chief Ministers of the States concerned, but the dispute remained unresolved. Eventually, in July 1986, Tamil Nadu made a formal request to the Government of India under the Inter-State Water Disputes Act 1956 to set up a tribunal to resolve the dispute. For various reasons the Central Government did not immediately establish a tribunal; it continued to favour a negotiated settlement. Meanwhile, a long-pending petition by some Tamil Nadu farmers to the Supreme Court for an assurance of irrigation water from the Cauvery came up for a hearing, and the Supreme Court, taking note of the failure of negotiations and the fact that a request from Tamil Nadu for a tribunal was pending, ordered the Central Government to establish a tribunal within a month. The Government of India accordingly established the Cauvery Waters Tribunal on 2 June 1990.

After almost 12 years the Tribunal seems at last to be slowly nearing the end of its labours; it may be reasonably expected to give its final Award within the next six months, or at least in a year's time. (It gave an Interim Order in 1991 which, as we shall see shortly, gave rise to a secondary dispute between the contending States.)

The Adjudication Mechanism

Article 262 of the Constitution and the Inter-State Water Disputes Act 1956 (ISWD Act) enacted by Parliament under that Article are important components of Indian federalism: they provide for the adjudication of disputes between States over the waters of inter-State rivers. The manner in which that mechanism has been working, the changes that were felt to be needed and the recent amendments to the Act, have already been dealt with and need not be gone into here. Suffice it to say that in the Cauvery case the adjudication process has been running a troubled course.

The Nature of the Dispute

The essence of the Cauvery dispute is a conflict of interests between a downstream State (Tamil Nadu) which has a long history of irrigated agriculture and has in the process been making substantial use of Cauvery waters, and an upstream State (Karnataka) which was a late starter in irrigation development but has been making rapid progress and has the advantage of being an upper riparian with greater control over the waters. To this dispute Kerala (an upstream State with a relatively modest demand for Cauvery waters) and Pondicherry (the lowest riparian with a very small demand) have become parties. Any fair sharing would have to provide for the legitimate interests of all four parties. Unfortunately, driven by the forces of party politics, the Governments of the two principal contending States have over the years generated and fostered strong chauvinistic sentiments among the general public, which tend to limit their (the Governments') own negotiating freedom and flexibility. The Cauvery is a fabled river with strong historical, religious, and cultural associations in both Karnataka and Tamil Nadu. In both States the mention of the Cauvery evokes a strong emotional response. The dispute has become (or has been made to become) a major issue in electoral politics. In both States, all parties tend to take a strong stand on this issue, making it risky for whichever party is in power to give the impression of being weak or of failing to protect the interests of the State. That is the reason why this dispute has become so intractable.

The Interim Order of 1991

However, before we go further into the dispute, we must get out of the way the secondary dispute that arose from the Tribunal's Interim Order

(IO) of 1991, particularly because it was in that context that popular frenzy led to tragic violence and introduced a new strain in the relationship between Tamils and Kannadigas, adding one more potentially divisive element in a country already torn by dissensions of various kinds. Fortunately, that episode soon came to an end, and a fragile calm has prevailed during the past several years; but the situation is an unstable one and violence could easily erupt again.

The IO of 1991 was passed by the Tribunal in response to Tamil Nadu's plea that pending the completion of the adjudication process—which could take time—there was need for some assurance of water for irrigation in the Cauvery basin in the State. Initially the Tribunal took the position that it was not empowered to consider such an appeal for interim relief, but on an application by Tamil Nadu the Supreme Court ruled that the State's request for interim relief stood included in the terms of the Central Government's reference to the Tribunal, and that the Tribunal could therefore consider it. The Tribunal then considered Tamil Nadu's plea and passed the IO to the effect that Karnataka should ensure an annual release of 205 TMC of Cauvery waters to Tamil Nadu (of which 6 TMC should go to Pondicherry), and also laid down a detailed monthly schedule of releases. (The Tribunal arrived at the figure of 205 TMC by taking the average of the flows of 10 years from 1980–81 onwards, after eliminating abnormally good and bad years.) Tamil Nadu wanted the Central Government immediately to notify the IO in the Gazette (as would be done in the case of the final order), and ensure its implementation. On the other hand, Karnataka felt that the IO was unfair and unimplementable, and sought to nullify it by promulgating an Ordinance for protecting the interests of its farmers. Faced with this situation, the Central Government made a reference to the Supreme Court for an opinion on certain issues. The Supreme Court gave the opinion that the IO should be notified, and that the Karnataka Government's Ordinance was unconstitutional. The Central Government then published the IO in the Gazette, and proposed to set up a committee to monitor the implementation of the IO, but failed to do so because of strong opposition from Karnataka. Karnataka let its Ordinance lapse, but continued to feel that the IO was unimplementable. It made a reference back to the Tribunal, but the Tribunal reaffirmed its order, observing that situations of abnormally low flows could be dealt with when they arose, and that a *pro rata* adjustment could be made. However, it did not lay down a detailed formula for such contingencies.

Opinion in Karnataka was strongly against compliance with the IO, and the Government took no steps for its implementation. This caused no immediate practical difficulty because there were good rains for three successive years. However, Tamil Nadu was anxious that the binding nature of the IO should be recognized (particularly as this would have a bearing on the implementation of the final Award as and when passed), and moved the Supreme Court for a direction that the IO should be implemented.

Meanwhile, rainfall was inadequate in 1995–96 and real difficulties were apprehended. Tamil Nadu went to the Supreme Court asking for an immediate release of 30 TMC by Karnataka (calculated with reference to claimed shortfalls in releases) to save the standing crops in Tanjavur. The Supreme Court asked Tamil Nadu to approach the Tribunal with its request. The Tribunal heard Tamil Nadu's plea and Karnataka's objections, and then ordered Karnataka to release 11 TMC immediately. As Karnataka showed no signs of doing this, Tamil Nadu went back to the Supreme Court asking for a direction that the Tribunal's order for the immediate release of 11 TMC should be implemented. The Supreme Court chose not to issue any such direction (partly because it was shortly going to hear Tamil Nadu's pending plea for a direction regarding the implementation of the Tribunal's IO), and instead requested the Prime Minister to intervene and find a political solution by consensus, and failing that, to give his own decision regarding immediate relief. The Prime Minister gave his decision in favour of an immediate release of 6 TMC, and set up a committee to determine the water needs for saving the standing crops in both the States. After an all-party consultation, Karnataka released 6 TMC. The Committee set up by the Prime Minister submitted a report, but the report was not made public; however, its contents gradually became known. In any case, it was of no great significance, as its remit was limited.

Some confusion was undoubtedly created by the return to the political arena *after* an order by a judicial authority (the Tribunal) and the virtual supersession of that order by a decision of the Prime Minister. This raised questions regarding the status of the Tribunal's orders and the implications of non-compliance with those orders. It was hoped that those issues would be resolved when the Supreme Court finally heard and passed orders on Tamil Nadu's pending petition on the implementation of the IO.

The Cauvery River Authority

That never happened. At one stage, in response to a query by the Supreme Court, the Central Government stated that it proposed to establish a machinery (as provided for in Section 6A of the ISWD Act) to oversee the implementation of the IO. In pursuance of that indication, it circulated a draft notification in 1997 seeking to establish an official-level, professional-cum-bureaucratic Cauvery River Authority, but this was strongly objected to by Karnataka for many reasons, one of them being that the Authority would have been vested with far-reaching executive powers (including the takeover of the State reservoirs). Eventually, in 1998, a *political* (as distinguished from a professional or executive) Cauvery River Authority (CRA) was set up, with the Prime Minister heading it and the Chief Ministers of the four States (Karnataka, Tamil Nadu, Kerala and Pondicherry) as Members. This was conceived of essentially as a limited dispute-resolving mechanism in relation to the IO, with no wider functions of planning or management. This was underpinned by an official-level Monitoring Committee headed by the Union Water Resources Secretary which would determine the facts in respect of complaints of non-implementation of the IO and service the political body. Under the circumstances then prevailing (which we need not go into here) the Tamil Nadu Government accepted this arrangement and withdrew its pending petition from the Supreme Court. Thus the Supreme Court did not have to pronounce on the legal status of the Tribunal's decisions. (That question now stands resolved in a manner following the recent amendments to the ISWD Act.)

The political CRA has not in fact been of much significance. It meets very infrequently and has played no useful role so far. However, given the discord between Karnataka and Tamil Nadu, even the limited agreement implied in that problem-solving mechanism, imperfect and ineffective as it might seem, was a positive development and was generally welcomed.

Those developments in relation to the IO not only created confusion but also tended to distract attention from the main water-sharing dispute. An impression was created that a major breakthrough had occurred and that a longstanding dispute had been resolved. (Claims to that effect were made by some distinguished commentators, and formed the basis of a political advertisement issued by the Bharatiya Janata Party (BJP)). Not only is the vexed, long-pending Cauvery dispute still before the

Tribunal, but as mentioned earlier, the CRA has very little to show for itself even in relation to its limited functions in relation to the IO.

In any case, the IO was only of temporary significance, and will cease to be relevant once the Tribunal gives its final award.

The Main Dispute: Apprehensions and Hard Positions

Reverting to the main dispute, the apprehensions and grievances underlying the positions of the two principal disputants need to be taken note of.

Karnataka feels that its late start in irrigation development should not mean any curtailment of its rights to make the fullest possible use of Cauvery waters for agricultural and other development. It also has a grievance regarding the past: it is of the view that the interests of Mysore, a princely State, were subordinated to those of Madras, a part of British India, and that the 1924 agreement was an unfair one. That charge has however been questioned, and it has been argued by some in Tamil Nadu that the Government of India and the Secretary of State were even-handed between the two parties; that the 1924 agreement was the result of hard negotiations and was fair to both sides; and that it was welcomed by the then Dewan of Mysore. Be that as it may, the sense of grievance exists, and must be taken note of. However, in the years following that agreement, both sides proceeded to expand the area under irrigation and neither strictly conformed to the limits imposed by the agreement. Thus, even if Karnataka's grievance regarding the old agreement were justified, that grievance has perhaps not much relevance now. In any case, Karnataka has been able to construct a number of storages on the Cauvery system. As an upper riparian it tends to assume a primacy of rights over Cauvery waters, with only the residuary flows going to Tamil Nadu, thus virtually, if not explicitly, invoking the Harmon Doctrine (of territorial sovereignty, including sovereignty over the waters that flow through the territory).

Tamil Nadu, as the lower riparian, feels threatened because its long-established irrigated agriculture based on a substantial use of Cauvery waters, with a centuries-old history behind it, is now vulnerably dependent on diminished and diminishing flows as a result of upstream development. From time to time, in years of inadequate or poor rainfall, it finds itself obliged to seek releases of small amounts of water from Karnataka and to invoke the intercession of the Central Government for this purpose. It feels that this dependence on the goodwill and generosity of its upstream neighbour puts it in a precarious position; it would prefer to obtain a clear recognition of a legal *right* to a share in the Cauvery waters. It therefore tends to take a legalistic stand on past agreements and on the principle of prescriptive rights arising from prior appropriation.

Each State has thus taken a stand on what it considers all its rights: Karnataka asserts an unqualified right to use Cauvery waters for the benefit of its farmers, and Tamil Nadu keeps insisting on its right to historic flows and the permanence of the 1924 agreement. We have to get away from both those untenable propositions. Neither the Harmon Doctrine nor the doctrine of prescriptive rights has wide acceptance; the principle that stood enshrined in the old Helsinki Rules—and that was favoured by successive tribunals in this country—was that of equitable apportionment for beneficial uses. Tamil Nadu must realize that historic flows cannot be restored; that it cannot hold a veto on upstream development; that it must learn to live with reduced flows; and that it can do so through a combination of better water management, avoidance of waste, local conservation of rain water, conjunctive use of surface water and groundwater, changes in cropping patterns, and so on. Karnataka must recognize that Tamil Nadu is a co-riparian with a right to share in the waters of the common river and not a poor relative asking for charity; that long-established irrigated agriculture and the way of life built around it should not be unduly disrupted; and that the abstraction of water by the upstream State should not be done in such a manner as to cause serious difficulties ('substantial harm' in the language of the old Helsinki Rules or 'significant' adverse effects in the language of the 1997 UN Convention relating to international watercourses) to the downstream State. With that kind of wisdom on both sides there should be no difficulty in arriving at a fair and reasonable settlement, a settlement which would also take care of the needs of Kerala and Pondicherry.

What Needs to be Done?

Given the rigid positions of the two State Governments and the perceived risks of compromising (or seeming to compromise) on those positions, how is an agreement to be brought about? It was precisely that difficulty that led to the establishment of a Tribunal. There was a view that a negotiated agreement was impossible and that adjudication was the only way out. Whether that view was right or wrong, the Tribunal was in fact established and is now in the final stages of its work. It is a mechanism provided by the Constitution. The best course now would be to wait for and accept the results of the adjudication process.

There is no reason to fear that the Tribunal, which is a judicial body presided over by a distinguished judge, will be anything other than fair and objective. Karnataka had suggested earlier that the Tribunal's proceedings should be suspended until a set of 'guidelines' or a 'national policy' on the principles which should govern the sharing of the waters of inter-State rivers was formulated and adopted by the National Water Resources Council (NWRC), but the prospects of a national consensus being reached on water-sharing principles are far from bright. Nor is it clear whether the process of conflict-resolution will be greatly facilitated by any document that emerges from such an exercise: it is unlikely to set forth any principle other than that of equitable sharing which successive tribunals have been trying to apply, and it is bound to be couched in very general terms which will still need detailed elaboration and application in each case with reference to the facts and circumstances of the case. A process of negotiations or adjudication in each case seems inescapable. Fortunately, despite its earlier reservations Karnataka has in fact been participating actively in the proceedings of the Tribunal, and as mentioned above, it appears that we can expect the final Award of the Tribunal in the not too distant future.

At the same time, even at this late stage conciliation efforts need not be ruled out. If a group of eminent persons commanding respect in the two States could undertake such an effort, and is able to persuade the parties to arrive at an agreement, then the agreement can be reported to the Tribunal and converted into an Award, thus giving it a statutory backing.

However, whether a negotiated agreement is reached or the adjudication process is completed and an Award is received, the ground for its acceptance and implementation has to be prepared in advance. It is necessary to rescue public opinion from the miasma of error, confusion, prejudice and anger that has tended to cloud it, and create a climate which is favourable to the acceptance of a reasonable settlement, whether it results from conciliation or adjudication. This requires a campaign in both States by persons of goodwill. Efforts in this direction were initiated some years ago, but they seem to have petered out.

Will the final Award of the Tribunal, when received, mark the end of the dispute or the beginning of further trouble? Will good sense prevail, or will ill-informed public opinion and the short-sighted calculations of party politics render rationality impossible? We must hope that the former will be the case.

Postscript

As in 1995, difficulties arose this year (2002) too because of poor rains and low flows in the Cauvery. Tamil Nadu complained of failures on the part of Karnataka to release waters in accordance with the Interim Order, and of consequent distress in the Cauvery delta. The *kuruvai* crop was said to have been lost, and the uncertainty about water made it difficult to plant the *samba* crop. Karnataka, on the other hand, said that there was not enough water for the needs of the farmers of the State and that it was very difficult to spare water for Tamil Nadu. In the absence of a formula in the Interim Order for the sharing of waters in a bad year, the *impasse* could not be resolved.

It is not proposed to narrate here in detail the tortuous course of this wrangle back and forth between the CRA and the Supreme Court. What needs to be noted is that the Cauvery basin farmers of Karnataka, particularly in the Mysore–Mandya region, whether on their own or under leadership of some kind, adopted an extreme and intransigent position, and mounted an agitation (that tended to turn violent) against *any* release of waters to Tamil Nadu. They did not want the State Government to implement the IO of the Tribunal, or release 1.25 TMC every day as ordered as an interim measure by the Supreme Court, or accept even the reduced release of 0.8 TMC ordered by the CRA. Tempers rose and the relations between Kannadigas and Tamils came under renewed strain. The State Government found it difficult to deal with this situation, and preferred to placate the farmers rather than comply with the laws of the land and the directions of the Supreme Court. This virtually amounted to a breakdown of federalism, and a constitutional crisis seemed to be looming. The Chief Minister of Tamil Nadu, for her part, chose to voice the grievances of the State in a confrontationist rather than a conciliatory manner. The film stars in Tamil Nadu decided to hold protest meetings, go on fasts, and so on. One extremist group threatened to disrupt the supply of electricity to Karnataka.

Ultimately, the Supreme Court, in response to two 'contempt of court' petitions filed by Tamil Nadu, passed severe strictures on the Karnataka Government for defying its orders and the decisions of the CRA, and ordered the release of Cauvery waters to Tamil Nadu. At the moment, the Supreme Court is watching compliance by the Karnataka Government with these directions before pronouncing its final orders on the contempt issue. (This is an imaginative and innovative use by the Supreme Court of its powers for punishing contempt.) The Supreme Court also disapproved of the expression by the Chief Minister of Tamil Nadu of a lack of confidence in the CRA.

The crisis has for the present been defused, the waters seem to be flowing, the north-east monsoon has set in, and the Karnataka Government seems to be finally dealing with its agitating farmers. It seems possible to hope that there will be no recurrence of a flagrant violation of the laws of the land or an open defiance of the Supreme Court. However, it is difficult to say whether good sense has dawned on all, and whether the Final Order of the Tribunal, when received, will be smoothly implemented. One hopes that that Order will be a carefully considered one that takes all relevant factors and arguments into account, and will seem *prima facie* reasonable to all parties to the dispute. One further hopes that it will contain suitable provisions for dealing with the contingency of low flows. The ground has to be prepared in advance for the acceptance of that Order. The feelings of the Karnataka farmers still remain inflamed; those feelings have to be assuaged, misperceptions removed and a process of education in the acceptance of conflict-resolution mechanisms needs to be undertaken. Goodwill between Kannadigas and Tamils needs to be promoted. Some civil society initiatives in both States seem to be called for. (The author made a modest move in this direction with a small informal meeting at Bangalore on 27 October 2002. An account of those discussions has been given in Appendix II.)

In the long run, the root cause of the conflict, namely, an excessive draft on the waters of the Cauvery by the farmers in both the States, has to be tackled through economies in water-use, changes in cropping patterns and the integrated and sustainable management of water from all sources (canals, groundwater, rainwater, soil moisture, etc.).

4

The Story of the National Water Policy 1987

National Water Policy 1987: A Narrative

This chapter will provide a brief narrative account of how the idea of a national water policy originated in the 1980s, took shape and emerged finally in September 1987 as an official declaration, and will then proceed to offer a commentary. As the 1987 Policy has now been superseded by the 2002 Policy, this is purely of historical interest. It is offered partly as personal memoirs and partly as administrative history.

In 1985, fresh winds were blowing through the corridors of the Government of India. Those were the early months of Rajiv Gandhi as Prime Minister; along with some of his (then) close associates, he was trying to bring a corporate management approach to the task of running the Government of India. That approach had its limitations, but it was a refreshing change from the older style of functioning. At any rate, there was much excitement in the air at the time, and for a period (which, alas, did not last long) civil servants had a pleasant and stimulating environment in which to work.

It was in that ambience that I became Secretary to the Government of India, Ministry of Irrigation and Power, Department of Irrigation, on 1 July 1985. After some intensive reading and wide-ranging discussions with a number of persons both within and outside the government, I attempted an account of sectoral problems, weaknesses, policy issues and the points that needed attention in relation to water; and from that account I derived a statement of objectives and tasks. The first item in that statement of objectives and tasks was as follows:

Reorientation of Ministry: The Irrigation Ministry to be regarded as the Ministry of Water Resources Development and Management.

- Convene early meeting of National Water Resources Council.
- Prepare paper on National Water Policy.

I also put in a similar entry as the first item in the Action Plan prepared for the Ministry. Whether in response to my suggestion or (more probably) independently of it, the Department of Irrigation was made a separate Ministry of Water Resources late in September 1985 or early in October 1985. The way seemed clear for a reorientation of the Ministry towards issues of water policy. This required among other things a meeting of the National Water Resources Council (NWRC).

The NWRC was a prestigious body set up at the instance of the National Development Council (NDC), with a composition similar to that of the NDC: it had all the Chief Ministers of the States and the Lieutenant-Governors of the Union Territories as well as a number of Central Ministers as Members, and the Prime Minister as the Chairman. The Union Minister of Water Resources was the Vice-Chairman, and Secretary, Water Resources, was the Secretary of the NWRC. The NWRC was an important institution for making federalism operative in relation to water, but though it had been constituted by a Government Resolution in 1983, it had not met even once by the middle of 1985. One of the first things to be done, therefore, was to convene a meeting of the NWRC. At my request, the first meeting of the NWRC was scheduled for 30 October 1985. As the agenda for that meeting, in my capacity as Secretary of the Council I prepared and circulated a brief document entitled 'Towards a National Water Policy: Issues for Consideration', based on the study and discussions referred to earlier.

The first meeting of the NWRC was very well attended and was a major event. As the speeches were proceeding, the Prime Minister turned to me and asked me in an aside: 'You wanted a meeting, and here it is; you have circulated an issues paper: what do you want to get out of this meeting? What would you like me to do?' I respectfully suggested that it would be a good idea to set up a group of Ministers to formulate a draft national water policy for consideration and adoption by the NWRC. He nodded and started putting down some names on a memo pad and then pushed it acrosss to me with a questioning glance. I noted that there were many non-Congress Chief Ministers in it. The Prime Minister smiled, and said that making them participate in the formulation of the policy document was the best means of achieving a consensus and ensuring a smooth passage for the document in the NWRC.

While diverse and wide-ranging views were expressed in the statements of the Chief Ministers, there was complete agreement on the urgent need for the formulation of a national water policy.

The composition of the Group of Ministers set up by the NWRC with a mandate to prepare a draft national water policy document for its consideration was as follows: Union Minister of Water Resources as Chairman, seven Chief Ministers of States (Andhra Pradesh, Arunachal Pradesh, Karnataka, Punjab, Rajasthan, Tamil Nadu and Uttar Pradesh) and four Central Ministers (Agriculture, Energy, Transport and Environment) as Members, and Secretary, Water Resources as Secretary. The Group was to complete its work in six months, but in fact it took three more months. The Group held five meetings in all. It considered an outline paper or checklist entitled 'Elements of a National Water Policy Document' circulated by the Secretariat. Two other papers were also circulated by the Secretariat: a paper on the constitutional/legal framework relating to water resources, and a paper on groundwater development and utilization. The issues enumerated in the checklist were discussed at great length, and valuable contributions were made by many members.

In between these meetings, a number of economists, engineers, administrators, social scientists and other eminent persons were consulted on the formulation of a national water policy at a meeting held at the Ministry of Water Resources on 18 March 1986; a meeting of the Parliamentary Consultative Committee of the Ministry was devoted exclusively to the consideration of the same subject; some policy issues also came up during debates and discussions relating to the Ministry in Parliament; and the issues were further discussed at the Regional and National Conferences of Irrigation and Water Resources the Ministers of States convened at the instance of the Union Water Resources Minister. In the light of all those discussions, and after internal consultations with the officers of the Ministry and the Central Water Commission, and with the Minister, I prepared a draft national water policy paper. At the penultimate stage, the draft policy paper was also circulated to several Central Ministries, and the comments received from them, as well as those received from the State Governments, Members of Parliament, MLAs, experts and other eminent persons, were placed before the Group of Ministers set up by the NWRC. The draft was discussed thoroughly in the Group, some amendments were made, and finally, at its fifth meeting held on 5 August 1986, the Group approved the draft national water policy for submission to the NWRC.

For considering the draft policy document, a meeting of NWRC was scheduled for 5 February 1987. Unfortunately, the meeting was postponed. I left the Ministry in May 1987 and moved on to another assignment. The NWRC finally met in September 1987 and approved the draft (piloted by my successor, Naresh Chandra) with a few minor amendments. Thus, in September 1987 the National Water Policy came into being.

(It had been my intention to submit to the NWRC, along with the draft national water policy document, an operational statement showing, with reference to each paragraph, what needed to be done, by which agency or agencies and within what time frame. Unfortunately, that statement was evidently lost sight of after I left the Ministry, and what went to the NWRC was only the draft national water policy document.)

A Commentary

Having regard to the importance of water, the federal structure of this country and the nature of the allocation of responsibilities in respect of water in the Constitution, the need for a national consensus on a policy framework was clear. It was an awareness of this that led to the formulation of the National Water Policy (NWP) 1987. Behind that exercise lay the recognition that there was a need to move away from an excessive preoccupation with discrete projects and towards issues of resource policy. The aim was to get all the States to subscribe in broad terms to a minimal set of propositions of a general nature, which could then form an agreed basis for more detailed policy-making and action plans. The NWP 1987 as adopted by the NWRC represented such a national consensus resulting from extensive discussions. Over the years it became a basic reference document and was frequently cited in discussions relating to diverse issues of water resources development and management.

The NWP 1987 was a slim document and its structure and contents were clear at a glance. The following (abridged) selection of a few important propositions from it may serve to give the reader a rough idea of its contents: water is a scarce and precious national resource; the basis of planning has to be a hydrological unit, such as a basin or sub-basin; project planning should be for multiple benefits, based on an integrated and multidisciplinary approach, with special regard to the human, environmental and ecological aspects; groundwater exploitation should be regulated with reference to recharge possibilities and considerations of social equity; the conjunctive use of surface water and groundwater should be ensured; in water allocation, the first priority should be for drinking water; there should be a close integration of water-use and land-use policies, and the distribution of water should be with due regard to equity and social justice; water rates should cover maintenance and operational charges and part of the fixed cost; farmers should be progressively involved in the management of irrigation systems, and the assistance of voluntary agencies should be enlisted in this context. There were also sections relating to flood-prone and drought-affected areas.

A consensus on a document of this kind was not arrived at without compromises between divergent points of view. A few instances may be of interest. In paragraph 1.8 of the document, the categorical statement about water being a national resource stands qualified by a reference to the needs of the States. Similarly, the document envisages planning for a river basin as a whole but makes no specific reference to river basin commissions or authorities, because of apprehensions on the part of the State Governments that their own powers might be eroded by the formation of such bodies; there is merely a vague reference to 'appropriate organizations' (paragraph 3.3). Further, again in paragraph 3.3, a reference to the optimal use of available water resources stands qualified by the words 'having regard to subsisting agreements or awards of tribunals.' These qualifications and compromises need not be regarded as unfortunate as they served the purpose of facilitating a broad consensus.

Further, certain aspects or areas were not covered in the NWP 1987. For instance, it refrained from making any observations on the changes that might conceivably be needed in the legal/constitutional framework for the successful implementation of the propositions that it set forth. Here again, a divergence of views and apprehensions on the part of some States resulted in an avoidance of the subject. Similarly, the document was silent in regard to the question of the sharing of inter-State river waters and the resolution of disputes in relation to such sharing. If an attempt had been made to formulate a set of principles on this matter, the debate on those principles would have been protracted, and the entire policy document would have got bogged down in that controversy. In any case, there was already a well established conflict-resolution procedure in terms of Article 262 of the Constitution and the Inter-State Water Disputes Act 1956; and (at that stage) it seemed to be functioning reasonably well.

Three years after the adoption of the Policy by the NWRC in September 1987, the Government of India constituted an official-level body called the National Water Board (NWB) (with the Union Secretary, Water Resources as the Chairman and the State Secretaries as Members) to consider the modalities of implementation of the NWP 1987, and in general to function as a kind of supporting agency for the NWRC. It must be noted that the NWB was not a body set up *by* the NWRC, as a kind of executive arm or servicing agency; it was a body set up by the Ministry of Water Resources with the objective of assisting the NWRC. Despite its establishment and the several meetings which were held from time to time, the operationalization of the NWP 1987 did not make much headway; it continued to remain largely a set of general propositions, though some of the ideas set forth in it independently gained currency through the reports of various committees or through certain Plan programmes. Now, without ever bringing the 1987 Policy properly into effect, we have moved on to a new Policy document of 2002. This will be looked at in the next chapter.

5

The National Water Policy 2002

At the meeting of the National Water Resources Council (NWRC) held on 1 April 2002, the amendments proposed to the National Water Policy (NWP) 1987 were approved, and a new NWP 2002 came into being. This is a matter of some importance and the nature of the changes need to be considered very carefully, but first the perception of the need for a revision must be noted.

Perception of Need for Revision

We have seen how the recognition of the need for a national consensus on a policy framework on water led to the formulation of the NWP 1987, and how it remained largely a set of general, non-operationalized propositions.

Looking back on it now, an unsympathetic critic could say that the NWP 1987 was a good beginning but that it did not go far enough. It certainly took note of the emerging environmental and equity concerns, but perhaps not adequately; priority was accorded to drinking water, but this was no more than a pious declaration; and, despite the intention of shifting the focus from projects to resource policy issues, it still devoted what may now seem a disproportionate amount of space to large irrigation projects. It was a well meant but limited and inadequate start.

By the late 1990s there was a general recognition that it was necessary to review the NWP 1987 and make changes in it. The National Commission for Integrated Water Resource Development Plan referred to this in its Report of September 1999. It observed that it was clear that 'if the NWP were being drafted today, it would need to show a much greater awareness of the present climate of opinion in regard to many matters...'; then proceeded to outline some of those concerns, and added that '… the concerns and considerations outlined above seem to call for a fresh exercise of drafting a policy document rather than going in for amendments and additions'.

Revision: Flawed Process and Approach

During the late 1990s, an attempt to revise the NWP 1987 was undertaken, and a draft was placed before the NWRC at its fourth meeting held on 7 July 2000. The draft failed to be approved because of reservations of diverse kinds on the part of different State Governments; it was then referred to a group of Ministers, Central and State. That group held several meetings and completed its work, and the new document that emerged was placed before a meeting of the NWRC on 1 April 2002, and was duly approved.

Unfortunately, this was a wholly internal governmental exercise. Suggestions that the draft amendments under consideration be put into the public domain, and that a series of broad-based meetings involving all concerned be held at various places in the country for wide-ranging discussions, were made to the Ministry of Water Resources, but elicited no response. The usual secretiveness of governments, and their reluctance to share documents or hold consultations with people and institutions outside, seems to have prevailed.

Unfortunately, again, what went on behind closed doors within the governmental system was not a root and branch examination but an 'amendment' exercise, and the resulting document can hardly be described as a 'new policy'. The NWP 1987 may have been inadequate, incomplete and sketchy, but it at least had a structure and flow. The 'revision' exercise played havoc with that structure and flow, without achieving any significant purposes. A valuable opportunity for fresh, careful and fundamental thinking has been lost. It took 15 years to revise the NWP 1987, and it may be another decade at least before the NWP 2002 is replaced by a new document. Meanwhile we have in place a policy document that can only be described as a 'non-event'.

National Water Policy 2002: Examining the *Fait Accompli*

However, we do have a new NWP 2002 in force, and faced with that *fait accompli*, we have to consider what sort of consequences, good and/or bad, are likely to flow from it. That involves three questions:

(*a*) (starting from the given text) what are the nature and implications of the changes (additions, amendments, omissions) that have been made in the 1987 document; (*b*) (looking at the matter from a broader perspective) whether, and if so, how adequately the new document responds to the changes in the climate of opinion from 1987 to 2002, and how well some of our current concerns are reflected in it; and (*c*) (related to that question) whether the document is likely to facilitate, or at any rate fail to hinder, certain undesirable developments. This threefold examination may result in some repetition, though an attempt will be made to minimize this.

Changes in the Text

Several changes have been made in the 1987 text of the NWP: additions, amendments, rewordings, rearrangement of matter and so on. Ignoring those that are self explanatory, minor or inconsequential, let us look at the changes that are, or seem to be, important.

(i) Paragraph 1.3: 'Water is part of a larger ecological system. Realising the importance and scarcity attached to the fresh water, it has to be treated as an essential environment for sustaining all life forms.' The first sentence is correct, but the second is a muddled statement: water is essential and therefore important, and not the other way round; it is indeed scarce, but scarcity is something separate from essentiality (the latter does not derive from the former!); water is a life-support *means* or *resource* and not an *'environment'*; and, finally, water is in fact essential, and not merely to be *treated* as such. Nevertheless, in so far as this (badly expressed) sentence shows an awareness of the essentiality, scarcity and importance of water, it is to be welcomed. But what follows from that recognition? What are its policy implications? The answer to that question fails to emerge in the rest of the document.

(ii) Paragraph 3.2: This spells out the various means through which more water resources can be made 'utilizable'. The enumeration includes rainwater harvesting, which is welcome, but there is no special recognition of its importance: it is lumped together with 'inter-basin transfers' and 'desalination of brackish or sea water'!

(iii) Paragraph 3.4: 'Watershed management through extensive soil conservation, catchment area treatment, preservation of forests and increasing the forest cover and the construction of check dams should be promoted. Efforts shall be to conserve the water in the catchment.' This is a welcome addition. (One wonders whether this was added after the Prime Minister's speech to the NWRC; please see comment (v) in the next sub-section.)

(iii) Paragraph 4.1: This is a new paragraph regarding institutional mechanisms, but while it calls for the appropriate reorientation/ reorganization/creation of institutions, it does not indicate what kind of reorientation etc., is required. The operational significance of this paragraph is not clear. Further, the last two sentences suddenly turn from the larger question of institutional reform and reorientation to the narrower (though important) one of the neglect of the maintenance of water resource schemes (because this is 'under non-plan budget'). This jump unwittingly diminishes the scope of the paragraph.

(iv) Paragraph 5: This is based on a paragraph in the 1987 document on priorities in water allocation, and is subject to the same weaknesses. In the first place, 'drinking water' needs to be expanded to include water for washing and cooking, and we should consider not only human needs but also those of livestock and wildlife. (A reference to animals does occur elsewhere.) Secondly, apart from an absolute priority to drinking water as so defined, there cannot really be any other *inter se* priorities among water uses: the relative priorities in each case will have to be determined with reference to the locational, socio-economic and other circumstances of the case. (In so far as the document recognizes this, the 'priorities' become meaningless.) Thirdly, 'irrigation' for commercial agriculture cannot be on the same footing as irrigation for sustenance or livelihood. The whole business of 'water allocation priorities' is an exercise of doubtful value.

(v) Paragraph 5: In the same paragraph again, there is a new entry. 'Ecology' is given fourth priority after drinking water, irrigation and hydropower! Water itself is subsumed in ecology (as recognized in paragraph 1.3), and sound ecological balance will determine the continued availability of water. Ecology, then, is anterior to all water-uses, and giving it fourth place shows a distorted sense of priorities. In fact, ecology cannot figure at all in a list of priorities for allocations of water: it is absurd to make an *allocation* of water for ecology. Ecological considerations may impose restraints on the various uses of water, and on the draft that we make on nature: ecology itself cannot be treated as being among the competing recipients of allocations of water.

(vi) Paragraph 6.6 says that the drainage system should be an integral part of project planning. This is an obvious statement, and hardly a matter of 'policy'. It forms part of the current guidelines on project preparation and need not have been elevated to a policy statement.

(vii) Paragraph 6.8: 'The involvement and participation of beneficiaries and other stakeholders should be encouraged right from the project planning stage itself.' Superficially that seems all right, but it is a revealing formulation in many ways. First, the expression 'beneficiaries and other stakeholders' gives the game away: the beneficiaries in the command area are evidently considered the primary 'stakeholders'. One has serious reservations on the concept of 'stakeholder' and has expressed them elsewhere; but if it is to be used, surely those who are displaced, who lose their land, habitat, occupations, livelihoods, and centuries-old access to natural resources—who are in fact *stake-losers*—ought to be regarded as the primary stakeholders. They cannot be put on the same footing as, much less ranked lower than, the beneficiaries in the command area on whom the project confers water rights that they did not have before, and who are therefore *stake-gainers*. Secondly, consider the word 'encouraged'. Who is to do the encouraging? The government, presumably. In other words the 'participation' envisaged here is a condescension by the government which will retain the primary role in planning, and will graciously 'consult' the people concerned. This is in the same old tradition of 'top-down' planning, and not true participation. A proper approach would give the people a primary, active role: they will then be co-actors and not mere 'participants' in a government-initiated and government-managed process. If that is in fact the intention, one can only say that it is not brought out by the above-quoted sentence, which gives the impression of paying lip-service to an idea that is currently in fashion. Thirdly, the 'participation' referred to here is in the context of 'projects'; this has nothing to do with the idea of local community management of common pool resources; we shall return to this point later.

(viii) Paragraph 9.5: This new paragraph stresses the need to 'get optimal productivity per unit of water'. This is a welcome addition.

 (ix) Paragraph 9.6: This addition refers to the 'reclamation of water-logged/saline affected lands', but says nothing about how this problem is to be avoided in the future. (However, there is an indirect reference to this in paragraph 7.3 which talks about the conjunctive use of surface and ground water.)

  (x) Paragraph 10: (*a*) Starts with a bland affirmation of the need for 'storages' with no indication that there has been a protracted and complex debate on the subject.

   (*b*) The paragraph goes on to advocate a 'skeletal' national re-settlement and rehabilitation policy. A draft rehabilitation policy has been under consideration for over 15 years and is nowhere near final adoption. Should not the NWP 2002 have had something to say on the factors that have stalled the draft and the urgent need to overcome them? (The continuing divergence on this is reflected in the word 'skeletal': the national consensus is to be restricted to a 'skeletal' level!)

   (*c*) The exhortation to ensure that 'construction and rehabili-tation activities proceed simultaneously and smoothly' is yet another bland formulation that ignores the controversy over the *pari passu* condition. This was initially intended to mean that construction should proceed *pari passu* with (i.e., not proceed ahead of) activities on the environmental and reha-bilitation fronts, but was subsequently taken to mean the opposite, i.e., that there was plenty of time for environmental and rehabilitation measures in view of the long periods in-volved in project construction, and that the former could proceed gradually along with the latter. It was that under-standing, which weakened the needed sense of urgency, that came in for severe criticism from several quarters; it now stands enshrined in the NWP 2002.

 (xi) Paragraph 12: Recommends a participatory approach to water resources management, but is open to the same criticism as that made in point (vii) in relation to paragraph 6.8. Consider the wording: '... by involving not only the various governmental agencies but also the users and other stakeholders ...'. Government comes first; people are *also* to be involved, and they figure primarily as users or stakeholders, not as owners. (Incidentally, this paragraph seems to contain the only reference to women in the document.)

(xii)   Paragraph 13: This is an addition, and an important one at that. It calls for the encouragement of 'private sector participation in the planning, development and management of water resources projects', on the ground that this may 'introduce innovative ideas, generate financial resources, introduce corporate management and improve service efficiency', etc. Some may applaud, some others may deplore and yet others may be doubtful about this approach. If the NWRC wants to subscribe to these propositions, it is entitled to do so, but is it entitled to present them to the nation as 'policy' without a discussion? If the draft document had been made public, and if there had been a series of debates on the subject in various parts of the country with diversified participation, the formulation might have been challenged and undergone changes.

(xiii)   Paragraph 14.3: This introduces the idea of 'minimum flow' in streams, and is welcome in so far as some flow is better than no flow, i.e., total abstraction leaving a dry bed. However, there is a danger here. To those who regard water flowing in the stream as 'wasted' and only water abstracted as 'used', the idea of a 'minimum flow' may carry the sanction for the obverse, i.e., 'maximum abstraction'. We ought to be concerned, *not* with the *minimum* that should be grudgingly allowed to flow, but with what is needed for maintaining the integrity of the river regime. What needs to be minimized is interference with the natural regime.

(xiv)   Paragraph 20.2: The reference to the need to 'monitor and evaluate the performance and socio economic impact of the project' is welcome. One wishes that a specific reference had been made to the re-evaluation of the social, economic, environmental and other impacts of a project after a few years of operation so as to redetermine the soundness of the original investment decision; however, the wording as it stands can perhaps be utilized for that purpose.

(xv)   Paragraph 21.1: In view of the absence of agreement among the states on water-sharing principles, this is merely a general exhortation. (The present writer is sceptical about the usefulness of attempting a statement of inter-State water-sharing principles, but that point has been argued elsewhere.)

(xvi) Paragraph 21.2: This calls for amendments to the Inter-State Water Disputes Act 1956. The amendments have been already made, and need to be separately examined. This has been done in an earlier chapter.

(xvii) Paragraph 22: Begins bravely by calling for a 'paradigm shift' in the management of water resources, but ends tamely by advocating a mere re-prioritization of financial allocations 'to ensure that the needs for development as well as operation and maintenance of the facilities are met'.

(xviii) Paragraph 25: This adds to the areas for research, but gives unmistakable evidence that the dominance of the engineering discipline continues.

The foregoing selectively covers the textual changes from the 1987 document to the 2002 one. Let us now proceed to consider the extent to which the new NWP reflects the changing climate of opinion.

National Water Policy 2002 and Current Concerns

Some deficiencies of the NWP 1987 were noticed earlier: an inadequate reorientation towards resource policy issues, a continuing preoccupation with 'projects', an inadequate appreciation of   environmental and equity concerns and so on. It is a matter of regret that the NWP 2002 marks no great improvement in regard to these points and that the old weaknesses continue.

(i) For instance, from the beginning to the end of the document, references to 'projects' occur over and over again. The major preoccupation of the NWP 2002 is in fact 'water resource projects', i.e., dams and reservoirs. 'Water resource development' is implicitly understood to mean big projects, and the entire policy document has been structured around that understanding. There is no explicit recognition of the extent to which that kind of thinking has come in for questioning in recent years, and the kinds of alternative approaches that have been urged for consideration. Whatever final policy position the document wished to adopt, should there not have been some recognition of one of the most important controversies of our time? Once again, if the draft document had been thrown open to public debate, the orthodoxy of the Establishment would have been challenged and it would have been necessary to take a diversity of views into account. Perhaps that was the reason for the avoidance of public debate.

(ii) In regard to 'environment', 'ecology', 'sustainable development', etc., there are certainly a few more references to these in the NWP 2002 than in the old document, but what significance do they have? Words and phrases have been added here and there in the old text because these ideas are in the air, but one looks in vain for a transformation in the ways of thinking. Environmental and 'sustainability' considerations are referred to in paragraphs 1.4, 3.3, 4.1, 6.2, 6.3 and 6.4, and such references are welcome, even if one suspects that they have not been deeply thought through. However, what is one to make of the last sentence in paragraph 6.3? 'The adverse impact, if any, on the environment should be minimised and should be offset by adequate compensatory measures. The project should, nevertheless, be sustainable'. One gets the impression of a gesture in deference to 'correctness'.

(iii) Again, the words 'multisectoral', 'multidisciplinary', etc., occur more than once, but not 'interdisciplinary' or 'holistic'. Perhaps too much should not be read into this, but there is some ground for wondering whether there is an inadequacy or deficiency in thinking here.

(iv) An important change in the climate of opinion since the 1980s is the new stress on people's participation. We have already seen that the NWP 2002 shows only a limited understanding of this (see the comments on paragraphs 6.8 and 12 in the earlier subchapter 'Changes in the Text'. We shall return to this shortly.

(v) The two most egregious failures of the NWP 2002 are in relation to water harvesting and community management of common pool resources. Water harvesting is a theme that has been much discussed in recent years. By now it is unmistakably clear that this will have to be a significant component of our water resource planning in the future. And yet, the NWP 2002 shows no appreciation of this. Speaking subject to correction, the only reference it makes to rainwater harvesting is in paragraph 3.2, where (as already mentioned) it figures as one of many ways of increasing the quantum of 'utilizable' water resources along with several others, with no indication of the importance attached to it. The Prime Minister's speech to the NWRC included the following remarks: 'Let this meeting of the Council send out a powerful message that harnessing every drop of rainwater is a national priority. We should lay special emphasis on localized, decentralized harnessing of water resources....' Unfortunately, there is nothing corresponding to this in the NWP 2002, unless we treat paragraph 3.4 as being related to this.

(vi) Similarly, the idea of community management of common pool resources, now widely accepted, finds no place in the NWP. Here, again, there is no congruence between the Prime Minister's speech and the NWP. The Prime Minister had said: 'The policy should recognize that the community is the rightful custodian of water. Exclusive control by the government machinery ... cannot help us to make the paradigm shift to participative, essentially local management of water resources.' There is nothing corresponding to this in the NWP. As already mentioned, its references to 'participation', 'Water Users' Associations', 'stakeholders', etc., are only in the context of big projects. The idea of community management does not figure at all in the document. The only reference to 'community' occurs in the last paragraph: 'Concerns of the community needs to be taken into account in water resources development and management'. This is hardly what one means by 'community management'. Interestingly enough, it appears that even this weak sentence was absent in the draft placed before the NWRC, and seems to have been hastily and grudgingly added after the Prime Minister's speech. (Incidentally, the words 'paradigm shift' used by the Prime Minister actually occur in the NWP, but in a different context, and there too the initial force is later weakened, as pointed out in the comment on paragraph 22 in the preceding sub-section 'Changes in the Text'.)

In passing, it may be mentioned that the Guidelines on Participatory Watershed Development issued by the Rural Development Ministry in 1994 was an enlightened document that sought to promote people's participation in the true sense. It appears that those who drafted the NWP 2002 made no use of it.

(vii) Related to water harvesting, watershed development and community management is the question of the relationship between the state and civil society, and that involves in turn the question of the relationship between formal law and customary law. Difficult issues in this regard have actually arisen. While the Madhya Pradesh Government has been actively trying to promote the role of civil society and facilitate a constructive, co-operative relationship between it and the state, the Irrigation Minister of Rajasthan has chosen to take a confrontationist position vis-à-vis civil society. These are extremely important issues, but the NWP 2002 takes no note of them.

(viii) Again, the major constitutional change in the form of the 73rd and 74th amendments came after the NWP 1987. The implications of these for future water management need to be taken note of. The NWP 2002 has nothing to say on the subject.

The document is in fact essentially the result of a tug-of-war between the Central Ministry of Water Resources and the corresponding Departments at the State level, with the former trying to enlarge its role and the latter trying to whittle it down. The latter seem to have had some success, considering the fact that the national rehabilitation policy is to be 'skeletal' (paragraph 10), and that the scope and powers of the river basin organizations are to be decided by the basin States themselves (paragraph 4.2). Does this mean that the Centre has been persuaded to abandon its role under Entry 56 in the Union List? Be that as it may, the Centre-State tug-of-war seems to have left no scope for considering the role and functions of the newly created third tier in the federal structure. That tier evidently played no part in the formulation of the NWP 2002.

Hidden Dangers?

The failures and deficiencies of the NWP 2002 represent not merely lost opportunities, but also potential impediments to the kinds of thinking and action that are called for. Moreover, there are elements in it that are fraught with danger. The doctrinaire advocates of water markets and of the treatment of water as a tradable commodity may be able to cite paragraph 13 of the NWP relating to private sector participation in support. In all fairness, it must be said that the document does not actually describe water as a tradable commodity. (The word 'asset' in the very first line of the document has been questioned by Rajendra Singh and others as implying a perception of water as a commodity, but that is perhaps a rather far-fetched interpretation; the stress there is on the adjective 'national', and in any case in later paragraphs the word 'resource' is used.) However, considering the important controversy on the subject of water as 'commodity' and water as 'commons' (or 'a basic right' or 'a sacred resource')[1], the NWP should have attempted an explicit formulation. Similarly, water markets serve some useful purposes but can also do great harm. Here again the absence of a well considered formulation on this subject is a serious weakness of the NWP 2002.

A reference must be made in this context to a disturbing development. It has been pointed out by Vandana Shiva and others that the view of water as a tradable commodity now stands enshrined in some international agreements; that a handful of giant corporations are trying to get control over the world's water resources; that they may receive the full protection of the World Trade Organisation (WTO) system; that attempts by member countries trying to preserve their natural resources from the onslaughts of these corporations may fall foul of the WTO regime; and that there is a danger of developing countries losing control over their own natural resources, including water. There may be a difference of views on that perception, but this is surely a matter of considerable importance. There is no recognition of this in the NWP 2002, and in the event of the dangers pointed out by Vandana Shiva materializing, no help can be derived from this document.

Conclusion

One regrets to have to say that the NWP 2002 is a patchwork quilt uncharacterized by cogency or coherence, and uninformed by a philosophy or vision. It is indeed a great pity that a valuable opportunity for a comprehensive review and a well considered new statement has been lost. Having said that, however, we must recognize that the NWP 2002 is now the officially declared policy, and will stay in place for some years. A wholly negative and dismissive approach to it will be inappropriate. What needs to be done is to adopt a practical, 'opportunistic' approach (using the latter term in a good sense). We must of course point out the deficiencies and persist in calling for revisions, but we must at the same

---

1 These matters are discussed in Chapters 7, 8 and 10, section II.

time try and do what we can with the document as it exists. It does talk about the environment, ecology, sustainability, equity, social justice, conservation, participation, role of women, involvement of stakeholders, and so on, and there is no reason why we should not invoke the aid of those sentences (however loosely, imperfectly or vaguely they may be worded) in our advocacy of the right causes, objectives and actions. We must also treat the Prime Minister's speech to the NWRC as an integral part of the NWP, and interpret the latter in the light of the former. Through proper, effective and repeated citation, the document can in fact be made to come alive and serve useful purposes. Attempts can also be made to ensure that the deficiencies in national policy are rectified to the extent possible in the State policies that are to follow. These are the ways in which some value can be salvaged from what would otherwise turn out to have been a wasted effort.

6

The Idea of River Basin Planning

The National Water Policy (NWP) 1987 stated that water-resource planning should be done with reference to a hydrological unit such as a basin or sub-basin. The idea of 'basin planning' has found wide acceptance, and frequently it is expanded to 'sustainable basin planning'. That phrase brings together two different ideas, namely, 'sustainability' and 'basin planning'. There are questions to be raised about both those notions, and moreover, neither term necessarily implies or entails the other. However, before going into those issues, let us take a quick look at India's 'experience' of basin planning and management. The inverted commas are intended to express a doubt whether there has in fact been much experience.

As early as 1956, the Union Parliament passed the River Boards Act (RBA) under Entry 56 in the Union List, but this provides only for advisory boards and not for river basin authorities vested with powers of management. In fact, no river boards, even of an advisory kind, have been set up under this Act. The Act has virtually remained a dead letter. When the NWP was being formulated in 1985–87, the question of river basin authorities came up but most States were apprehensive of their own powers being eroded,[2] and eventually the NWP made only a vague reference to 'appropriate organizations'.

The Damodar Valley Corporation (DVC) was indeed intended to be a river basin authority, but that effort, modelled on the lines of the Tennessee Valley Authority, was prior to the adoption of a quasi-federal Constitution. While it has served some useful purposes, it has not in fact functioned as a river valley authority. Its multiple functions (flood

---

2 This statement is based on the author's personal knowledge as Secretary of the NWRC and its Group of Ministers in 1985–87.

moderation, power generation, irrigation and the general development of the area) were whittled down over the years, and the DVC is today mainly a power-generating body, and much of that power is, ironically enough, thermal power. (Incidentally, the DVC is administratively under the Ministry of Power and not under the Ministry of Water Resources.) Similarly, the Bhakra–Beas Management Board (also under the Ministry of Power) is a system-management body and not a basin planning organization. The Betwa River Board was set up by a separate enactment, but this was only for the specific purpose of overseeing a particular project. The Brahmaputra Board was another board set up under a specific Parliamentary enactment. It was vested with powers of execution of projects, but its role has been confined largely to the preparation of a master plan and the formulation of a few large projects (none of which has made any headway); it has not, and perhaps could not have, grown into a river basin authority. There are various organizations set up under Government Resolutions. Some, such as the Bansagar Control Board, were meant to supervise specific projects. The Ganga Flood Control Commission was limited to the preparation of master plans for flood control. The Narmada Control Authority, a body set up under the orders of the Narmada Waters Dispute Tribunal with limited functions relating to cost allocations and the rehabilitation of project-affected persons, was later enlarged to cover the monitoring of environmental aspects, but it is not in the nature of a Narmada Basin Authority. (Nor can that description apply to The Narmada Planning Group and the Narmada Valley Development Authority, which are internal organizations of the Gujarat and Madhya Pradesh Governments respectively). The Krishna Waters Dispute Tribunal envisaged the establishment of a Krishna River Authority, but this was in 'Scheme B', which was not part of the Award itself and was therefore not operative. In regard to the Cauvery river, attempts to establish a professional, empowered Cauvery River Authority failed, and instead, a political body (with the Prime Minister of India as Chairman and the Chief Ministers of the basin States as Members) for dealing with conflicts relating to the implementation of the Tribunal's Interim Order of 1991 was set up in 1998. This is not a basin planning or management body, and even for the limited purposes in view it has not been a notably successful arrangement.

There are no river basin authorities or boards in India, of the kind that exist in France or Holland. As mentioned earlier, the NWP 1987 did talk about planning for a hydrological unit such as a basin or

MAP 6.1: *River Basins of India*

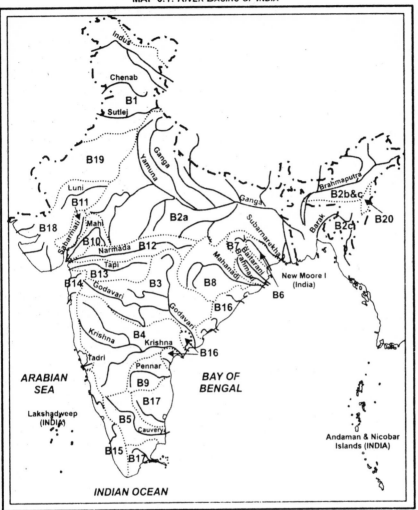

ARABIAN SEA

BAY OF BENGAL

INDIAN OCEAN

Lakshadweep (INDIA)

Andaman & Nicobar Islands (INDIA)

New Moore I (India)

NAME OF THE LINKS

B1	Indus Basin	B12	Narmada Basin
B2 a,b,c	Ganga, Brahmaputra and Barak (Meghna) sub-basins (upto international border)	B13	Tapi Basin
		B14	West-flowing rivers from Tapi to Tadri
B3	Godavari Basin	B15	West-flowing rivers from Tadri to Kanyakumari
B4	Krishna Basin	B16	East-flowing rivers between Mahanadi and
B5	Cauvery Basin		Godavari and between Krishna and Pennar
B6	Subarnarekha Basin	B17	East-flowing rivers from Pennar to Cauvery and
B7	Baitarani–Brahmani Basin		from Cauvery to Kanyakumari
B8	Mahanadi Basin	B18	Luni Basin
B9	Pennar Basin	B19	Area of inland drainage in the Rajasthan Desert
B10	Mahi Basin	B20	Minor river basins draining into Bangladesh
B11	Sabarmati Basin		and Burma

Source: Central Water Commision

MAP 6.1 RIVER BASINS OF INDIA

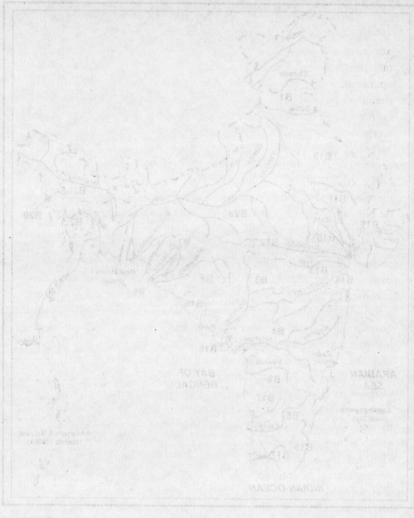

sub-basin and about 'appropriate organizations' for the purpose, but this (like most of the general statements in the NWP) was never operationalized. The National Commission for Integrated Water Resources Development Plan (1999) recommended river basin organizations of a *representative* kind (in the hope that this would prove more acceptable), with a very large principal body of a 'general assembly' or 'river parliament' kind and a smaller (but still not very small) executive committee.[3] However, these ideas have not made much headway. In the NWP 2002 adopted on 1 April 2002, river basin organizations are mentioned, but their scope and powers have been left to the basin States. As pointed out in the preceding chapter, this represents some loss of ground by the Central Government.

In the light of the foregoing, it may not be unfair to say that water resource planning in India has by and large tended to proceed on the basis of discrete, individual projects. That statement may be questioned. It could be plausibly argued that 'basin planning' of a kind was indeed attempted in several cases: for instance, Bhakra Nangal, Sardar Sarovar, Gandhi Sagar, and so on, were not supposed to be 'stand-alone' projects, but were envisaged as parts of larger systems. There was also a basin-wise assessment of resources and identification of storage sites by the Central Water and Power Commission in the 1950s. However, these were instances of 'basin planning' only in a limited sense, and even these were not wholly successful. 'Integrated planning' often means no more than planning a cluster of projects. In the 1980s, a multidisciplinary group in the Central Water Commission prepared a Ganga Basin Master Plan, but for certain reasons it was never made public, and under the circumstances, no comment can be made on its contents. The National Water Development Agency has been making assessments of basin surpluses and deficits, but this is essentially in the context of 'inter-basin transfers', for which it has been identifying possibilities of transfers, storages and links. This is not really 'basin planning'.

A truly integrated, holistic planning for a basin or a sub-basin would involve *interdisciplinary* planning for the basin or sub-basin, marrying

---

3 The author was a member of the National Commission for Integrated Water Resources Development Plan. He has some reservations on the workability of the recommendations referred to here, and has expressed them in a Note to the Report.

land-use and water-use, harmonizing diverse water uses on the demand side and integrating *all* 'development'—from local rainwater harvesting and micro-watershed development to 'mega' projects (and surface water and groundwater)—on the supply side, while at the same time fully internalizing environmental, ecological, human and social concerns, and fully associating the people concerned ('stakeholders') at all stages. That kind of basin planning has not really been seriously attempted in India.

We must now revert to the remark made at the beginning of this chapter that 'there are questions to be raised' regarding the key terms. Leaving 'sustainability' for later discussion, let us proceed to consider the second term, namely, 'basin planning', more closely. It derives from hydrology and is essentially an engineer's language. How did this idea originate? It arose from a recognition of the limitations and dangers of isolated project planning. Engineers built a dam here, a barrage there, a run-of-the-river scheme in a third place, flood control embankments in a fourth and so on. By experience they learnt that this was not the best thing to do, and that in planning any such intervention in a river they needed to take into account the river as a whole. That logic cannot be faulted. Discrete project planning is undoubtedly inferior to project planning within a larger framework. However, this is still a very limited vision for two reasons.

The first is that while widening our vision from a particular point on the river to the river as a whole we are still thinking only of the river, and not of the ecological system of which it is a part—by which it is sustained and which it in turn sustains. A basin is primarily a hydrological concept, not an ecological one. In fact, even from a hydrological point of view, a river basin approach suffers from the limitation that the boundaries of the basin may not coincide with those of groundwater aquifers. In theory everyone recognizes that water in all its forms—rivers, lakes and other surface-water bodies, wetlands, groundwater aquifers, atmospheric moisture, precipitation in the form of rain or snow, glaciers, and so on—constitutes a unity; in practice, however, basin planning is often focussed only on the river, ignoring, or taking only perfunctory note of, other forms of water.

Secondly, we are thinking of the river in terms of 'planning'; and if we ask 'What are we planning?', the answer is clear: 'projects'. As the discrete and fragmented planning of projects is unsatisfactory we wish to plan in a larger context, but we are still thinking in terms of projects.

We want to 'harness' the water resources of a river for human use through the application of science and technology, and it is in that context that the idea of basin planning emerges: the engineer wants to build better and larger projects. When he thinks of the basin as a whole, he thinks in terms of what from an engineering or economic point of view might seem 'optimal' locations. For instance, techno-economic optimality might suggest the concentration of agriculture in one part of the basin, industry in another, power generation in a third and so on. There could be some merit in such an approach, but it is essentially a centralizing tendency.

Thus, the idea of 'basin planning and management', which *prima facie* seems eminently sound, contains within itself the seeds of centralization and gigantism. We need to be aware of and on our guard against such tendencies. (The issue of gigantism, i.e, the controversy regarding big projects, will be discussed further in a later chapter.)

Subject to that caution, it is certainly necessary to take a comprehensive view of a river system as a whole. The initiatives that are taken at the micro-watershed level have eventually to be built into a harmonious, holistic, integrated basin-wide (or sub-basin-wide) total picture. Contrariwise, a broad basin-wide master plan can provide pointers to local initiatives. Unfortunately, such a comprehensive, holistic view of a river system requires institutional arrangements of a kind that have not so far been found feasible in our quasi-federal political structure.

After the above was written, the Report of the National Commission to Review the Working of the Constitution has become available. The Commission has recommended that the RBA 1956 be repealed and replaced by 'another comprehensive enactment under Entry 56 of List I'. The Commission goes on to say: 'The new enactment should clearly define the constitution of the River Boards and their jurisdiction so as to regulate, develop and control all inter-State rivers keeping intact the adjudicated and the recognized rights of the States through which the inter-State river passes and their inhabitants. While enacting the legislation, national interest should be the paramount consideration as inter-State rivers are "material resources" of the community and are national assets. Such enactment should be passed by Parliament after having effective and meaningful consultation with all the State Governments.' With all respect to the distinguished Commission, it is difficult to see much point in this recommendation. The RBA, which provided only for advisory boards, has remained a dead letter because of resistance by the States: how is a more comprehensive legislation going to be passed; and if passed, how is it going to be more effective? How are we to keep the States' rights intact, and at the same time ensure that the national interest is paramount? What are the operative implications of saying that 'inter-State rivers are "material resources" of the community and are national assets'? How does this statement impinge on the States' perceptions of their rights and their sovereignty? The Commission has not dealt with any of the difficulties that have impeded action in the past, nor has it tried to visualize in any detail how river boards or other forms of basin organizations acceptable to the Centre and to the States can be constituted. Perhaps it is not fair to expect a Constitution Review Commission to enter into such details, but without that kind of work no useful recommendations can be made. No wonder the Commission has been reduced to offering bland generalities which state conflicting propositions without reconciling them, and which have no practical value whatsoever, while at the same time appearing to be making radical recommendations for the repeal of an existing Act. With regret, one can only describe this as fuzziness.

II

Water: Perceptions, Rights, Laws

7

Perceptions of Water

Preliminary

In writing or speaking about water, it is easy to get muddled because of its multiple aspects or dimensions. Water is perceived by different people (or by the same people in different contexts) in different ways: as a *commodity*, as *commons*, as a *basic right* and as a sacred resource or *divinity*. Often when we are under the strong influence of one perception, other perceptions seem quite wrong. For instance, those who regard water as 'commons' or a 'common pool resource' tend to deny vehemently that it is a commodity. Contrariwise, those who see water as a commodity are often blind to the other dimensions of water. The truth is that we can say many things about water and be right. Commodity, commons, basic right, divinity: all these are partial perceptions; all are valid; we need all of them to understand the roles that water plays in our lives. We need to be aware that what is true of one of the multiple dimensions or aspects of water may not hold for another. Unfortunately, at any given time, one or more partial perceptions tend to dominate our thinking, and thus leads us into drawing wrong conclusions and formulating wrong prescriptions.

Water as Commodity

To many of us the notion of water as a commodity seems unacceptable. However, let us consider the following cases: (*a*) the use of water for irrigation in commercial agriculture; (*b*) the use of water for cooling or steam generation or industrial processes; (*c*) luxury uses by the affluent (in saunas, swimming pools, gardening and so on). Would a description of water as a commodity in such contexts be wholly inappropriate? Again, a hotel may need large quantities of water for keeping operational its kitchens, bathrooms, toilets, laundry, swimming pools and other recreational facilities, and may enter into contracts with supplying agencies for bulk water supplies on a regular basis; or a farmer may buy the irrigation water that he or she needs from another farmer in the neighbourhood: in these transactions is water not a commodity? Questions of equity, social justice and resource conservation do arise in such cases, and we may wish to limit or regulate or discourage the use, or ensure that proper prices are charged; but it seems hardly possible to rule out such uses or the related transactions altogether, and to the extent that they take place, water is indeed a commodity in these contexts.

However, no one is quite comfortable with the crude description of water as a commodity, so a more sophisticated formulation has been found: water is now generally described as 'an economic and social good'. The implications of this economic perspective are discussed further in the next chapter. Meanwhile, let us note that 'good' here does not necessarily have a connotation of moral approval; it is more akin to 'goods', i.e., objects of merchandise. We are not really too far away from 'commodity' here. That is not necessarily a wrong perception in some contexts, as illustrated above. However, it is indeed a wrong perception in other contexts such as water as life-support or as a common pool resource of a village community.

Water as Commons

The view of water as 'commons' or as a 'common pool resource' (CPR) is strongly advocated and is attractive, but two points need to be noted. The first is that the notion of 'commons' (as distinguished from private ownership) is of easy application in the context of a small lake or pond or tank or other waterbody on common land; we can think of it as owned by the community. With larger waterbodies, and with streams and rivers, difficulties begin to arise in the form of 'upstream *versus* downstream' issues, riparian rights and so on. However, we can still argue that the water source belongs to the community as a whole, or to 'civil society', and that the conflicts that arise can be resolved within that overall framework (though that benign formulation tends to breakdown when rivers cross national boundaries or even political divisions within a country). The notion of commons also runs into difficulties in the context of urban water supply systems (where an agency, whether public or private, supplies water to citizens by a network of pipelines from its storages), or in that of the supply of irrigation water through canals from large reservoirs, whether state-owned or privately owned.

The notion of commons has a value even in such contexts; what we are trying to do is to deny the private or state ownership of water and to vest that ownership in 'civil society'. But does any person, body or institution—even civil society—*own* water? Bypassing that question for the present, we can talk about 'community *management* of CPRs'.

Incidentally, we must be wary of unduly enlarging the geographical scope of the ideas of 'commons' and 'community'. If we widen the notion of 'community' to cover the State as a whole, or go one step further and encompass the entire nation under that rubric—and we might wish to do so for certain legitimate purposes—it may be difficult to resist further expansion to the globe as a whole. There are serious implications to accepting a description of water as a 'global commons', a natural resource that belongs to all humanity. The dangers are obvious enough. When we use expressions such as 'commons' or 'community management' we usually have a local context in mind (one village or a cluster of villages constituting a watershed), and it would be better to confine ourselves to that context.

The second point is that the community is a collectivity. The idea of community management of CPRs does not by itself imply any *individual rights* to water. The rights of individuals will be merely those that are agreed upon by the community (or conferred by civil society institutions) instead of being granted by the state or arising from contracts. The idea of individuals having rights to water is yet another perception.

Water as a Basic Right

Water sustains life. In that aspect, it is a basic need and therefore a basic right. This does not automatically follow from a description of water as 'commons'.

In the traditional societies of the past, people might not have needed the language of rights; customs and conventions would have been adequate. Indeed, some of those would have had the force of law: hence the expression 'customary law'. However, in the legalistic societies of today 'formal law' has become more important, and that is why it has become necessary to talk about the 'right to water'. We are talking about water as life-support, i.e., drinking water. As mentioned above, this is a basic or fundamental right, but is it also a 'human right'? This is discussed further in the following chapter and also in Chapter 10. Water as a basic right is a useful perception, but it has the potential of being asserted not only against the state (it is in that context that it comes to be formulated) but also against the community or civil society. We need both perceptions (water as commons and water as a basic right) and must learn to harmonize them.

Water as a Sacred Resource

The fact that water supports life, and that it is also a part of the natural environment, sustaining it and in turn being sustained by it, leads to its being regarded as 'sacred'. (This is not a reference to the ritual uses of water, though these reinforce the sacred aspect.)

In Indian tradition rivers have always been regarded as divinities—mostly feminine, though it must be noted that the Brahmaputra is regarded as masculine. The Hindu view of rivers as goddesses may be contrasted with the following passage from T.S. Eliot's *Four Quartets*:

> I do not know much about gods, but I think that the river
> Is a strong brown god—sullen, untamed and intractable,
> Patient to some degree, at first recognised as a frontier;
> Useful, untrustworthy, as a conveyor of commerce;
> Then only a problem confronting the builder of bridges.
> The problem once solved, the brown god is almost forgotten
> By the dwellers in cities—ever, however, implacable,
> Keeping his seasons and rages, destroyer, reminder
> Of what men choose to forget. Unhonoured, unpropitiated
> By worshippers of the machine, but waiting, watching and waiting.

It is interesting that in setting forth and criticizing a particular view of the relationship between humanity and nature, namely, the adversarial view that has been the dominant Western one[4] in the past, Eliot envisions

---

4 See the reference to the difference between the Western legend of Prometheus and the Indian legend of Bhagiratha in the discussion of the environmental dilemma in Chapter 25

rivers as gods, though (unlike in Indian tradition) he thinks of them as masculine.

The different perceptions that have been described briefly influence, but are not identical with, the different perspectives on water and related rights discussed in the next chapter.

Note

The caution about treating water as a 'global commons' expressed earlier in this chapter arises from certain apprehensions regarding the possible misuse of that idea by the richer and more powerful nations of the world as well as by transnational corporate giants in furtherance of their own interests. The idea of a World Water Contract proposed by Ricardo Petrella in *The Water Manifesto* (Zed Books, London, and Books for Change, Bangalore, 2001) may seem to obviate such apprehensions; but the feasibility of bringing about such a contract and keeping it safe from being hijacked by powerful forces needs careful consideration.

8

Water and Rights: Some Partial Perspectives

Questions of rights relating to, or in the context of, water resources arise in diverse ways and from different perspectives. Most of these are partial perspectives. This chapter will set these forth briefly in a sequence that has no particular significance, and then postulate, or at any rate raise the question of, the possibility of bringing them together in some kind of integrating framework or structure. (There may be overlaps among the different perspectives mentioned, and some concerns may be susceptible of being looked at from alternative perspectives.)

Riparian Perspectives

The riparian perspective is essentially one of rights to the waters of a flowing river inhering in, or claimed by, different users located alongside (or in the vicinity) of that river. This can arise at the level of households, farms, communities, villages or towns, but occurs in a more marked form at the level of political or administrative units within a country, or at that of co-riparian countries. Various principles and doctrines have been advanced in this regard: the Harmon Doctrine of territorial sovereignty, the rights of 'prior appropriation' or 'prescriptive rights', the Helsinki principle of 'equitable apportionment for beneficial uses', and, finally, the 1997 UN Convention on the non-navigational uses of international water courses, which has not yet been (and may fail to be) ratified by the required number of countries.

Good illustrations of riparian perspectives will be found in the inter-State Cauvery dispute in India (see Chapter 3, section I) and in the issues relating to India's water treaties with its neighbours: the Indus Treaty of 1960 (India–Pakistan), the Mahakali Treaty of 1996 (India–Nepal) and the Ganges Treaty of 1996 (India–Bangladesh), which are dealt with in

Chapter 19, section V. In the present chapter we are not concerned with the details of those disputes or the issues involved; what we need to look at is the form that 'resolution' tends to take. What generally happens in such cases is an *allocation* of waters among the different riparians (whether countries or units within a country), with each party receiving an allocated share to be used as it sees fit. It is possible to speculate whether, ideally speaking, a better result could not have been achieved through a joint, cooperative, integrated planning and management of the river as a system instead of dividing it up into fragments to be managed separately; and whether (as has been argued) a sharing of *benefits* may not be better than a sharing of *water*. However, agreement or even adjudication is better than dispute and discord, and if the ideal is not feasible, then the second-best solution should be welcome. The reader's attention is merely being drawn to a certain perceived limitation in the riparian perspective. Another limitation of the riparian perspective is that it tends to focus exclusively on the river and ignores everything else: groundwater aquifers, land, the ecological system of which the river is a part. The riparian perspective cannot be dismissed as irrelevant or unimportant; it will continue to have its place, but its limitations need to be kept in mind.

Federalist Perspectives

By 'federalist perspectives' one means the distribution of rights and powers in relation to water between different levels in the federal structure. This subject has already been partly discussed in Chapter 1, section I. The question that we need to ask in the current context is: what kind of rights are involved in this federalist perspective? The distribution of subjects into three lists (Union, State and Concurrent) in the Constitution is essentially a distribution of legislative power, i.e., the power to make laws. To those legislative powers of the Union Parliament and the State Legislatures correspond the executive powers of the Central and State Governments. We are therefore talking about the rights of *governments*. At a seminar on federalism held some years ago in Delhi, a discussant responded to an elaborate exposition of the constitutional provisions relating to water (the Entries in the State and Union Lists, Article 262, and so on) by saying that all this was merely a question of the sharing of powers between two bureaucracies, Central and State, and that she was more interested in the rights and powers of the people. That deliberately provocative statement contains a partial truth that needs to be pondered.

With the 73rd and 74th Amendments to the Constitution we now have a third tier in the federal structure besides the Centre and the States, namely the local level, i.e., villages and cities. One of the subjects to be devolved to that level is water management. This reform of democratic decentralization is as yet in the early stages and has not become a full-blown reality. As and when village panchayats become well-established institutions of self-government and powers and finances are devolved to them, they will begin to play an important role in relation to water. However, it needs to be noted that even with decentralization, while the state may come closer to the people, it will not become one with the people: the question of the relationship between the state and civil society will continue to be important.

Formal Law Perspectives

'Formal law' here includes judicial determinations. What exactly is the formal law perspective on water in India? It is difficult to give a simple answer to that question because what prevails is a confused and complex position. The late Chattrapati Singh, in his masterly first chapter entitled 'Water Rights in India' in the book on *Water Law in India* edited by him (Chattrapati Singh 1992), gives an account of the change from the traditional view of water as a 'natural right' in pre-colonial times, through early colonial laws which continued to recognize and incorporate that view to some extent while bringing water and other natural resources within the control of the colonial rulers, to the final position of the full assertion of the state's eminent domain. He goes on to point out that in recent years, interpretations and rulings by the Courts, in widening the ambit of the fundamental right to life, as also in affirming citizens' rights to clean air and water, have tended to bring back to some extent the view of water as a natural right. Be that as it may, when we look at the present legal position, we find the following:

(i) The right to water, as a part of the right to life, is a fundamental right. (This is the outcome of judicial determinations.)

(ii) The right to clean air and water is also a fundamental right. (This is implied in the Environment Protection Act and the Water Pollution Control Act, and rulings by the Courts in environmental cases.)

(iii) To some extent the community's—particularly tribal communities'—rights of access to the natural resource base have also been
recognized in the Tribal Self-Rule Act (Provisions of the Panchayats
[Extension to the Scheduled Areas] Act, 1996) as well as certain
judicial decisions.

(iv) At the same time, the colonial legacy of the eminent domain of
the state, as laid down in the Irrigation Acts of the various States,
is not merely alive but is reasserted from time to time (as for
instance by the Rajasthan Irrigation Minister some time ago.) It
is also occasionally confirmed by Court decisions in certain cases.

(v) It is not clear whether legislative (and, correspondingly, executive) powers over water as determined by the constitutional
provisions, and the eminent domain conferred by the Irrigation
Acts, imply the *ownership* of rivers and other surface waters by
the state. Governments tend to assume this. This has been questioned in some Tribunal Awards. However, in most discussions
of a non-specialist kind there is a tendency to treat 'control' and
'ownership' as being synonymous.

(vi) A related issue is the following. In the context of the Inter-State
Water Disputes Act 1956, 'inter-State' really means *inter governmental,* and the question arises whether, in the event of two
Governments agreeing on a project on an inter-State river (or a
Tribunal laying down the details of such a project in its Award),
the rights of the affected *people* to be consulted about, or to
question, the project get extinguished. This question had come
to the fore in the Narmada (Sardar Sarovar) case, but the judgement of the Supreme Court, which was a strong affirmation of
the rights of governments, has to be taken as a dismissal of that
question.

(vii) International treaties or agreements over rivers (e.g., the Indus
Treaty, the Mahakali Treaty, the Ganga Treaty) are entered into
in exercise of the sovereign powers of the state (using that word
in the abstract sense, ignoring the Centre-State distinction).
Whether this implies ownership or not, it is clear that the state
does have the authority to bind the people to the treaty obligations.

(viii) In so far as groundwater is concerned, ownership rights over it
go with the ownership of the land under which it lies. (This is
elaborated upon in Chapter 9.)

All this does not seem to add up to a coherent whole. Here we merely take note of the existing position. In order not to break the continuity of this account of perspectives, the water law reforms recommended by Chattrapati Singh are discussed in a note at the end of this chapter.

Civil Society Perspectives

The community or civil society[5] perspective arises in three different but interconnected contexts: (a) efforts to protect people's rights, particularly those of poor, disadvantaged communities and tribal groups, vis-à-vis the state and its agencies in the context of large projects; (b) the move to revive traditional community-managed systems of water management that have gone into decline ('dying wisdom' in the language of the Centre for Science and Environment); and (c) new initiatives in social mobilization and transformation such as Anna Hazare's in Ralegan Siddhi in Maharashtra or Tarun Bharat Sangh's (Rajendra Singh's) in Alwar district in Rajasthan. Local leadership as well as NGOs play an important role in such initiatives, promoting an awareness of traditional systems and forgotten practices as well as of the people's rights under the old and new systems, kindling motivation, providing the necessary knowledge and skills, assisting in the resolution of conflicts, mediating between civil society and the state and empowering the people. The leaders and NGOs go beyond formal law and stress local traditions, time-honoured practices and conventions—in a word, 'customary law'.

We must recognize that formal law (as perceived and practised by the state and its institutions) and community initiatives (and the appeal to

---

5 The expression 'civil society' has come into vogue in recent years. It is widely used but rarely defined. In essence it means 'people' (at different levels in different contexts, ranging from local communities to the nation as a whole) conceived of as loosely organised in 'society' as distinguished from 'the state'. For that purpose 'society' should be adequate: why is the word 'civil' added? Perhaps this merely reinforces the distinction from 'state' or implies a distinction from 'political'; and perhaps the addition of the word 'civil' gives a certain resonance to the expression. One may also wonder whether there is an unrecognised normative component in the usage. When we refer to ourselves as members of 'civil society' we are unconsciously approving of ourselves, and excluding undesirable elements: no one would want to describe criminals or smugglers or corrupt people as members of civil society!

customary law and civil society institutions) do not go well together. The former is not only not hospitable to the latter, but is often positively hostile. In Rajasthan, after long-dry streams and wells had been regenerated by the people, the state stepped into claim control and the right to license fisheries. That particular dispute was somehow resolved for the time being, but the relationship between people's initiatives and the instrumentalities of the state is bound to remain an uneasy one. Again, we may commend the initiative of the people in establishing the Arwari Parliament (a body for resolving conflicts on the Arwari river in Rajasthan), but it has no statutory backing and can at any time be undermined by the state. In fact, community initiatives that are started with the best of intentions and for laudable purposes can unwittingly run counter to the formal law of the statute books. This has actually happened in Rajasthan, and the patched-up truce there is temporary.

'Participatory' and 'Stakeholder' Perspectives

In recent years even governments have begun to talk about people's participation. However, the notions governments have of participation, as exemplified in programmes such as participatory irrigation management (PIM), are generally rather limited. In the first place, participation is invited at a late stage in projects that are planned and implemented in a wholly non-participatory manner; secondly, it is often the inability of the state to manage a project and provide the planned services that leads to ideas of transferring responsibilities to the users; and thirdly, the state is usually unwilling to enter into a contractual relationship with the users and accept binding obligations with penalties for non-performance. The idea of participation is somewhat better understood in the context of the Joint Forest Management Programme and the Watershed Development Programme, and the 1994 Guidelines in respect of the latter are quite enlightened, though even here there are problems regarding implementation.

Reference must also be made here to the currently fashionable language of 'stakeholder consultation (or participation)'. This is part of the Dublin–Rio principles, and has gained greatly in currency in recent years. Unfortunately, both the terms of that phrase are dubious.

'Participation' can vary from the full involvement of the people from the earliest stages of planning (putting people at the centre) to the mere formality of asking for comments on a plan, programme or project prepared entirely within the governmental machinery, with no serious intentions of making any significant changes. (A similar point could be made with reference to the implementation stage as well.)

As for the term 'stakeholder', it is a flawed word that has great potential for misuse. First, it is a notion drawn by analogy from prospecting for oil or minerals and carries a connotation of an individualistic claim with an underlying implication of contestation. Secondly, it is an ethically neutral concept that lumps together every person or party having any kind of connection or concern with the project. Not only those who are likely to be adversely affected by the project or expect to enjoy the benefits that it will bring, but a wide range of others who are concerned with it in one form or another come within the ambit of the term. Thus, politicians, bureaucrats, engineers, donor agencies, consultants and contractors are all 'stakeholders'. The interests and concerns of these diverse categories may not in all cases be benign and legitimate, and some may have a more vital 'stake' than others, but the term 'stakeholder' makes no distinctions: it legitimizes and levels all kinds of 'stakeholding'. Everyone is a stakeholder, and the primacy of those whose land and habitat are taken away and who suffer a traumatic uprooting is not recognized by the term.

Even taking only two categories (in relation to a 'water resource development' project), namely, project-affected people and prospective beneficiaries, the vital difference between the two tends to get blurred by the bland assimilating term 'stakeholders'. There is a cruel irony in describing the involuntary and helpless victims of a project as 'stakeholders', and this is compounded when they are put on the same footing as those who stand to benefit from the project. Let us not forget that while in the case of the former *existing* rights (i.e., natural and often centuries-old rights of access and livelihoods) are taken away, in the case of the latter the project, by diverting river waters through canals, confers *new* rights not earlier enjoyed. The former are *stake-losers*, whereas the latter are *stake-gainers*.

A standard response to the hardships inflicted by such projects on the affected people is to say that while everything must be done to mitigate their hardship, development does involve costs and that some groups may have to accept a measure of hardship (sacrifice) in the larger interests of the nation. Without entering into a detailed discussion of the fallacies involved in this line of argument, let us merely note that 'sacrifice' is the wrong word to use for an involuntary displacement from land and homestead, and that the *imposition* of such a 'sacrifice' is morally indefensible. Mahatma Gandhi, in whose view the Benthamite doctrine of the maximum good of the greatest number was immoral, would surely have refused to countenance the 'sacrifice for development' argument. Nor is Pareto optimality an adequate answer to this: it is not enough to say that while some are enabled by state action to acquire wealth, others must be at least not worse off. That too is injustice. The project-affected persons (PAPs) must actually benefit from the project. However, while it is fashionable to refer to PAPs as 'partners in development', that sanctimonious formulation bears little resemblance to reality. Efforts to involve them in decision-making and to give them their rightful share in the benefits of the projects that impose hardships on them have either not been seriously pursued or have been unsuccessful.

Human Rights Perspective

In relation to water resources the human rights perspective is invoked in two ways: the right to water being regarded as a human right, and the resistance to displacement (for large projects) being formulated in the language of human rights.

The right to water is recognized as a part of the right to life and therefore a basic or fundamental right, but there has been some debate on whether it should be regarded as a 'human right'. A consensus that it should be so regarded seems to be emerging. That is too large and complex a subject to be fully discussed here, but three points need to be noted.

(i) The notion of water as a human right can be invoked only in relation to water as life-support, i.e., 'drinking water'. (This needs to be defined so as to include, within limits, water for cooking, bathing, washing clothes and personal hygiene, and also to cover water for livestock.) The right to drinking water so defined is undoubtedly a basic right, and can be regarded as a human right, but this cannot possibly apply to water for irrigation or industrial use, or for the generation of hydroelectric power. There can be no 'human right' to irrigation or hydroelectric power or indus-trial use of water[6]. As will be pointed out in a later chapter, the

---

6 A vague notion of 'human right to development' has indeed been mooted, but that kind of expansion of the idea of human rights will dilute it and rob

state in India has not been conspicuously successful in assuring drinking water to all, but by and large it would be true to say that in most societies in the world drinking water is not unduly problematic. It is irrigation that makes heavy demands on available water, and difficulties and conflicts arise largely in relation to water for irrigation. It follows that the formulation of a human right to water (i.e., drinking water) is only of limited significance, though it will undoubtedly serve some useful purposes. It is partly an assertion (often against the state) of freedom of access to the resource, and partly a demand (again, often addressed to the state) for provision, or at least an assurance of availability[7].

(ii) It is an *individual* right. It is not easy to bring the community's right to manage common pool resources within the rubric of human rights, though the case for *collective* human rights has also been argued.

(iii) As drinking water is only a small part of total water, the assertion of a human right to water cannot be in relation to the resource *per se*, but only in relation to the fulfillment of a certain need. In other words, the human right (if that is how we wish to describe it) is not to *water* as a natural resource but to the ability to achieve the quenching of thirst (and to perform the functions of cooking, bathing, etc.). That is an important distinction and not merely a quibble.

Let us turn now to the other route through which the idea of human rights comes into the picture in relation to water. Social activists who

---

it of its force. Further, it begs the whole question of what is meant by 'development'. Vandana Shiva might well ask whether we are arguing for a human right to maldevelopment!

7 In distinguishing between 'negative rights' (i.e., natural rights that are merely recognized by the state) and 'positive rights' (i.e., rights that are granted by the state), Chattrapati Singh says that the state has to *provide* the latter, whereas it has merely to *refrain from interfering with* the former (Chattrapati Singh 1992). That distinction in terms of provision is not wholly persuasive. It is not clear why a citizen cannot expect the state to see that he or she is able to enjoy a natural right. This surely goes beyond 'refraining from interference'. In a drought-prone area, does not the state have a responsibility to ensure the availability of drinking water, and if so, would that not include 'provision'?

are trying to bring about 'empowerment' of the people vis-à-vis the state, as also NGOs and individuals who take up the cause of people displaced or otherwise adversely affected by a project, or who seek to protect people from the high-handedness or callousness or violence of the agencies of the state, tend to invoke (among other things) the human rights perspective; but it is also a special perspective of lawyers, and there is a vast literature on the subject. It was a combination of NGO/activist concerns and lawyers' concerns (apart from the prevailing climate of international opinion) that lay behind the establishment of the National Human Rights Commission (NHRC). The 'tribal rights' perspective is a special variant of the human rights perspective. Resistance to certain state policies or actions that are perceived as being unjust is often articulated in the language of human rights. Unfortunately the response of the state to such movements is often one of incomprehension and force. The 'empowerment' activists in turn tend to postulate or assume an adversarial relationship between the state and civil society, and to fall into anti-state postures.

A Digression on Rights

In a world in which there is injustice, oppression, deprivation, distress and torture, the invocation of the idea of 'human rights' is both necessary and useful. Nevertheless, some reservations need to be entered here. The idea of 'rights' is central to modern thinking on social and political matters, but it has a negative side as well. It is often (but not necessarily, or always) an adversarial notion. Rights are usually *to* something, but they are also implicitly against something or someone. When one says 'This is *my* right', one is saying by implication 'It is not yours or anybody else's'; and if one says 'This is my *right*', one is asserting 'No one can take it away from me' or alternatively, 'It is somebody's duty to provide or ensure or protect it'. Consider the language surrounding the word rights: rights are asserted, claimed, demanded, defended, disputed, contested, fought for. Contestation and conflict seem pervasive in this language. Rights are of course necessary; it would be dangerous to downgrade or deprecate the notion; but one may well wonder whether it is not possible to think of an alternative language that stresses cooperation and harmony rather than conflict.

Turning to 'human rights', the first question that occurs to one is: why the qualification 'human'? The answer is that we want to say that certain rights are not a gift of the state but are anterior to the state; that they fundamentally inhere in us as human beings. Secondly, we want to recognize certain needs or linkages—food, habitat, access to a natural resource base—as basic to common humanity, and not to be taken away lightly or casually by the state or any other agency such as a corporate body. Thirdly, we want to stress human dignity: we do not want the state to treat anyone (not even hardened criminals) in a manner that demeans them, or subjects them to excessive and unreasonable hardship or pain, or undermines their sense of personality and self-respect. We are against brutality by the agencies of the state, cruel punishment, torture and so on. There can be no disagreement with any of this. However, it is one thing to assert the primacy or fundamental nature of certain rights, or to hold the state accountable for its behaviour, but quite another to say that these rights are anterior to the state. In the absence of a state what rights does a person have except what he or she can claim or maintain by physical force? The notion of 'rights' makes sense only in the context of a state or perhaps a civil society with its own sanctions. Some of those rights could of course be more basic than others. To take care of this, the concept of 'fundamental rights' would surely be adequate. If we do wish to use the term 'human', we are doing no more than stressing the primacy of certain rights; we cannot give them a validity independent of the state or civil society. Perhaps all of this is obvious, but it appears to the author that we do tend to use the term 'human rights' as if these rights had an autonomous existence.

A further problem with the term 'human rights' is the implied exclusion of the non-human. What about the rights of other species? The cruelty that humankind inflicts on other species is horrendous: do the latter not have any rights? An awareness of this has led to 'animal rights' movements, but what about the rights of the natural environment? Environmental rights, including the rights of a river to a minimum flow, have been recognized in court decisions in certain countries. What about the rights of planet Earth? How can humanity survive if planet Earth does not? Of course environmental rights and the rights of planet Earth can only be voiced by humans, and are being so voiced. The point that is being made here is that when we talk about human rights, we are implicitly divorcing humanity from the rest of nature.

As a means of avoiding both the undertone of conflict beneath the notion of 'rights' and the exclusiveness involved in ascribing rights only to human beings, an alternative notion of 'human obligations' or 'human responsibility' might be worth considering: obligations towards fellow human beings (whether they be co-riparians or anyone else, and including those whose habitat and ways of living we are about to disturb, in particular the poor and the disadvantaged); and obligations towards other species, nature and planet Earth. We shall return to this at the end of the chapter.

Environmentalist Perspective

This perspective stems from a concern to protect the natural environment from human depredations in pursuit of what goes by the name of 'development' (which Vandana Shiva would describe as 'maldevelopment'). That concern finds expression in the assertion of the rights of aquatic life, the rights of the river (for the maintenance of its integrity and regime) and the rights of the natural environment. Reference has already been made to this.

Economic Perspective

This perspective perceives water as an economic good and argues that it is best left to market forces. The thesis is that if property rights in water were defined and trading allowed, water markets would emerge, prices would be established, resource-conservation would take place, sustainability would be taken care of, equity would be ensured and conflicts would automatically be resolved by market forces. This is a very partial and limited perspective indeed. Recognizing this, the formulation about water being an economic good is usually modified to: 'water is an economic and a social good'. Water is certainly an 'economic good' when it is used for industry or agriculture, and perhaps a 'social good' when used for sanitation or in hospitals or for firefighting; but is even 'social good' an adequate description of water as a basic human and animal need (and indeed as the sustainer of the environment of which it is a part)? Can water in that basic aspect be reduced to a commodity like cement or steel or fertilizers or soap? Is it not more akin to the air? One is not ruling out water markets, which may have a role to play, but there are important issues of equity, social justice and sustainability that are unlikely to be the concerns of market forces. The glib answer to that will be that these can be taken care of through 'regulation', but regulation is far from easy.

Moreover, in the case of the other commodities properly so described the expression 'market' often refers to the global market. The argument is that everything need not be produced domestically, and that imports should be freely allowed. What application does this have to water? Water is not an internationally-traded commodity to an extent even remotely comparable to oil, and the availability of water in some part of the world is of no great relevance to a distant country or region suffering from water shortages. People need sources of water that are close to their homes and land; inter-country sharing or exchanges in relation to water can at best happen between neighbouring countries.

There is indeed a notion of trading in 'virtual water', i.e., in commodities, say, foodgrains, grown in water-abundant countries, but this has serious implications. A critical dependence on imports of foodgrains, even if the balance of payments position allows this, may introduce a degree of vulnerability that may not be desirable for a country like India. (There are of course other views on this subject, but that debate cannot be entered into here.)

The doctrinaire call for 'privatization' includes allowing the corporate private sector to build and operate dams across rivers for hydroelectric power and/or for irrigation. Assuming that the private sector is interested in investing in such capital-intensive, long-gestation, modest-return projects, how are the environmental and social impacts (which have presented serious difficulties to the state in past projects) going to be handled by the private entrepreneur and manager? Supply may match demand, but resource-conservation may receive scant consideration; resettlement and rehabilitation aspects are likely to be given grudging attention only to the extent that resistance by those affected and public opinion compel such attention; and it is naive to imagine that market forces will obviate conflicts or provide a magical route to their resolution. (This does not mean that one is arguing for a dominant role for the state, but merely that the alternative to the state is not necessarily the corporate sector.)

So far we have been considering some of the difficulties that arise from perceiving water as an economic good. A more basic question needs to be considered: can there be tradable property rights at all in relation to a natural resource such as water? Some doubts in this regard are set forth in the sub-section entitled 'Water Rights' in Chapter 10 ('Towards a National Water Law?') in this section.

Priorities Among Uses

There can be questions of relative priorities among the different uses at water: irrigation *versus* drinking water; rural *versus* urban demands; agricultural *versus* industrial demand; irrigation/power-generation *versus* flood-moderation; abstractions for use *versus* maintenance of minimum flows; etc. This is not one more 'perspective' on rights, but a question of socio-politico-economic choices. However, when conflicts arise and decisions are given by the Courts, this gets translated into the language of rights.

Conclusion: Towards a Total Perspective

That was a broad and somewhat sketchy outline of diverse perspectives. In each case, attention was drawn to some significant limitations. As stated at the outset, all of these are partial perspectives. That does not imply a questioning of their validity or relevance. All these perspectives are required, and each embodies important principles or values. What we need to avoid is the error of elevating a relative truth or value into an absolute one, or assuming that the partial perspective that one is adopting for the nonce is in fact a total or all-embracing one.

Can we assign centrality to one or more of these perspectives? Can we arrange all of them in a hierarchy? Can we integrate them into a harmonious whole? A hierarchy can be forthwith ruled out: the perspectives are too diverse and the inter-relationships (ranging from the tenuous to the close) are often too subtle and complex to lend themselves to being arranged into a hierarchical structure or order. Even the assignment of centrality to some perspectives is problematic, though access to water as a basic life-support resource and respect for the source of that substance, namely nature (planet Earth), seem to have an arguable case for such assignment. Keeping that in view, we have to integrate and harmonize the various perspectives as inter-linked and ineluctable parts of one all-embracing perspective. In that effort, perhaps a recourse to the rich, multi-faceted Indian notion of *dharma*—as an overarching all-embracing moral order—may be found useful. (The term *dharma* has several other meanings, including duty, responsibility, quintessential or defining function or avocation, etc.) If we think of diverse collocations such as:

- men/women;
- humankind/other species/nature;
- consumption/conservation;
- present/future generations;
- individuals/civil society/state;
- formal law/customary law;
- upper/lower riparians (including governments at different levels or of different countries);
- different users of water;
- ancient wisdom/modern science;

and so on, and ask in each case what is the *dharma* (or obligation or responsibility) of one element or component to the others, we may be able to bring all the perspectives together into one harmonious whole, which will be *Dharma* in the overarching sense. However, this is merely a tentative idea or philosophy that needs to be worked out carefully and in detail. All that is being offered here is an adumbration of an alternative to the usual approaches and formulations. As a non-lawyer the author offers it in fear and trembling, but is also emboldened by the thought that the point is not entirely a legal one.

Addendum: A Note on Chattrapati Singh's Recommendations

In parenthesis, we need to take note of the late Chattrapati Singh's recommendations for legal reform. He argues that the original natural rights over rivers and other natural waters belong to the people of India, and not to the government or the state; that people have a natural or fundamental right over what is essential for their life and which inherently belongs to them; that governments can have only a legal usufructuary right, and with the consent of the people; that when the government acquires any usufructuary right for specific public use it would have to compensate the original users or beneficiaries and define the 'public' in terms of all bearers of rights; and that to make the state accountable and to make water-use equitable for all in this nation, a number of amendments are required in the Easement Act, the Irrigation laws, *panchayat* and Municipal Corporation laws, Water Supply acts and other laws related to water (Chattrapati Singh 1992). (That summary is largely in his own words.)

There can be no disagreement with the broad propositions about recognizing people's natural rights, allowing the state only a limited role and making it accountable, compensating the original users when their rights are affected and so on. Similar points are made in several chapters in this book. However, the formulation cited above bristles with difficulties.

First, consider the words: 'and which inherently belongs to them'. What belongs to them? The right is to life and that which supports it, namely, drinking water. Access to drinking water is a fundamental right. Does it follow that 'rivers and other natural waters belong to the people'? Just as the state's legislative powers and powers of treaty-making do not necessarily imply the *ownership* of rivers, so too the natural right of all people to water as life-support does not necessarily imply ownership.

Secondly, the fundamental right of all people (in the hills or plains, in urban or rural areas, rich or poor, tribal or non-tribal) to drinking water, and the right of access of certain communities or settlements (particularly aboriginal or tribal communities) to the natural-resource base (forests, rivers or streams or water bodies, hillsides and so on) from which they have drawn their sustenance for centuries, are two different things. In the former case, the recognition of a fundamental right may imply a responsibility for provision or assurance of availability.[8] In the latter case, the state has to refrain from disrupting the age-old links between the community and its natural environment. This is where the idea of 'consent' comes in. As we shall see in another chapter, the World Commission on Dams has recommended that such disruption in the context of large projects must be based on 'free, informed, prior consent'. However, it is one thing to argue that people likely to be affected by projects must be consulted and their consent obtained before they are so affected, and quite another to advance a theory that the 'people' (does that mean the people of India as a whole?) must consent to whatever the state does. In fairness, it is not clear that Chattrapati Singh means this; but his formulation seems to carry this implication. On behalf of the Governments, Central and State, it will doubtless be argued that such consent was given when 'we, the people of India' gave to ourselves a Constitution, and that it is that Constitution which confers legislative (and executive) powers on the Central and State Governments in accordance with the three lists. What we *can* argue is that those Governments

8 See footnote 7.

must be held accountable to the people for what they do or fail to do, and that the idea of 'public purpose' that they readily invoke for interfering with people's rights must be more rigorously defined. That is presumably what Singh is in fact arguing.

The difficulty is that we have taken over in its entirety the colonial legal legacy, and that the Constitution itself is largely based on the Government of India Act 1935. Into that system we are trying to introduce elements, some traditional and some modern, that do not easily fit in with it or with one another. We have to reconcile the *individual* fundamental right of all people to water as life-support with the community's right of managing common pool resources, and both of these with the responsibilities (and therefore the rights) of the state to control, regulate, legislate, and so on. That complex task is not facilitated by Singh's formulation.

In this context, it is not quite clear why Chattrapati Singh refers to the rights of the state as 'usufructuary'. Presumably he means 'temporary rights of control or use over property belonging to another' (to use the definition given in the Oxford English Dictionary). But does the state *use* the water? It might, in some cases; but generally, it deals with the use of waters by the people. What the state enjoys is the power of provision, regulation, control, distribution and so on, and not (generally speaking) that of *using* the waters that it deals with. (It may of course derive revenues from its actions.)

Further, if 'usufructuary' means 'having temporary rights over property belonging to another', the question arises: does property in water belong to the people? We have already seen that the fundamental right to water does not imply ownership. Take the Cauvery for instance; does it belong to the people? Which people? Those of Kerala or Karnataka or Tamil Nadu or Pondicherry, or all of them? The answer is that it belongs to none of them, but all have rights of use. Similarly, the Ganga does not belong to the people of Uttar Pradesh or Bihar or West Bengal— or Nepal or Bangladesh—but all of them have rights of use over the waters of the Ganga.

Finally, granting that natural rights over rivers, whether that means ownership or use-right or 'sovereignty' (Singh questions whether the state has sovereignty over rivers), vest not in the state but in the people, what does this mean in operational terms? How will the people exercise those rights? Through their elected representatives at the three levels envisaged in the Constitution, or through a different set of institutional arrangements? Through watershed committees? Through NGOs? What will be the relationship between these institutions and the institutions of state? These questions will not be gone into here; the intention is merely to draw attention to the various perplexities that arise from Chattrapati Singh's formulation cited above.

Postscript

The author has considered whether any changes are called for in his discussion of human rights in this chapter in the light of what Upendra Baxi has to say in his recent book *The Future of Human Rights* (Oxford, 2002), but has decided to leave the chapter unchanged. Nevertheless, he would like to draw the reader's attention to that important book which is a work of formidable scholarship as well as of acute perceptions and insights. A deep and passionate commitment to the idea of human rights blazes forth from its pages.

9

Groundwater Legislation

Importance of Groundwater

The National Commission for Integrated Water Resources Development Plan, in its report of September 1999, puts the national groundwater resources at 432 billion cubic metres (BCM), and the utilizable component at 396 BCM. The importance of groundwater in our national life is evident: around 50 per cent of irrigated agriculture is based on groundwater, and 85 per cent of rural drinking water comes from groundwater. Even after all the major and medium irrigation projects (under construction or contemplated) are implemented, a substantial part of irrigation (not far below 50 per cent) will still depend on groundwater.

Problems Encountered

Before proceeding to legal questions, it may be useful to attempt a brief encapsulation of the problems that have emerged in relation to groundwater. There has been over-extraction (mining) of groundwater leading to depletion in several areas, and salinity ingress in some coastal zones. The National Water Policy (NWP) 1987 laid down the principle that extraction should not exceed the annual rate of recharge, but this has not been enforced. Similarly, the NWP 1987 talked about the conjunctive use of surface water and groundwater, but this too has not been translated into action. As a result, in some areas there is a situation of rising water tables and the emergence of waterlogging and salinity. A 1991 Report of a Working Group of the Ministry of Water Resources estimated the extent of waterlogged land in the country at 2.46 million hectares (MHA), and that of salt-affected land at 3.30 MHA. There are serious equity issues in the context of the power-driven extraction of groundwater through tubewells and borewells. There are also serious problems of pollution/contamination of groundwater. It is generally agreed that we have not made much headway in dealing with these matters. There are several reasons for this, social, political, economic and legal. The present chapter is concerned only with the legal aspects.

Constitutional Position

There is no specific reference to groundwater *per se* in the Constitution, though we could of course assume that 'water' includes groundwater. As mentioned in Chapter 1, section I the two basic entries relating to water, viz., Entry 17 in the State List and Entry 56 of the Union List, give the impression that the Constitution-makers were thinking primarily of river waters. Again Article 262 and the Inter-State Water Disputes Act 1956 (ISWD Act) passed by Parliament under that Article are about *river* water disputes. The possibility of disputes about underground aquifers that straddle State boundaries does not seem to have occurred to the Constitution-makers. Further, the Tribunals set up under the ISWD Act have refrained from taking an overall view of water resources covering both surface water and groundwater.

However, in the absence of a specific mention, it has generally been assumed that the reference to 'water' in the Constitution includes groundwater. Again, in the Fundamental Rights Section, the right to life has been held by judicial interpretation to include the right to water as a life-sustaining resource, and here too, 'water' must be taken to include groundwater, which is in fact a very important source of drinking water.

Central Groundwater Board: Basis?

On that understanding, groundwater seems to be a State subject falling within the purview of Entry 17 in the State List; and even the proviso relating to Entry 56 in the Union List (which is about inter-State rivers) has no relevance in the context of groundwater. How then do we explain the existence of Central organizations such as the Central Groundwater Board (CGWB)? It is of course partly a historical legacy. The Geological Survey of India (GSI), which was in existence long before the Constitution came into being, had a separate groundwater division. However, it was the Exploratory Tubewells Organization (ETO) which, with the addition of the groundwater division of the GSI to it, eventually became the CGWB; and when the ETO was set up in the 1950s, the Constitution was already in place. When the Central Government set up the ETO, did it specifically consider the Constitutional position, and if so, from what provisions or articles did it derive the authority to set up such an organization in the Central sphere? The historical accounts that the author has seen do not throw any light on this. It is of course possible that as only resource studies and explorations were being undertaken, and no executive actions were contemplated, the question of constitutional powers did not seem very important and never came up for consideration.

Central Groundwater Authority: Curious Position

If we turn to the Central Groundwater Authority (CGWA), the story is quite different. The fact that 'water' was a State subject was not forgotten, but it was bypassed. It was in a public interest case relating to an 'environmental' concern, i.e., the feared depletion of groundwater, that the Supreme Court mooted the establishment of a regulatory authority, and despite the doubts expressed by the Ministry of Water Resources, gave the directions that resulted in the establishment of the CGWA in 1997. It was the Ministry of Environment and Forests that set it up, and it did so under the provisions of the Environment (Protection) Act 1987.

The curious position that has come about is that the CGWA is the CGWB under another name, as the former has the same composition as the latter (with some additions); and while the CGWB is within the administrative purview of the Ministry of Water Resources, the CGWA is within the functional purview of the Ministry of Environment and Forests. In other words, what is virtually the same body functions in two different *personae* and capacities, under two different legal dispensations and within the areas of responsibility of two different Ministries. In theory, it is an independent, autonomous organization, but in practice it is acutely dependent on a constructive, supportive attitude on the part of the bureaucracies of two Ministries. It is not surprising that the CGWA, set up five years ago with much fanfare under the directions of an activist Supreme Court, has not as yet become very effective or even fully operational as a regulatory body.

Easement Rights

If we turn from the Constitution to the ordinary laws, we are struck by an asymmetry between groundwater and surface water. In so far as surface water is concerned, it is possible to argue, though with some doubts and hesitations, that the law recognizes only use-rights and not ownership or proprietary rights over flowing water. At the same time, as we have seen, the Irrigation Acts of the various States vest control over river waters in the State Governments. (This 'eminent domain' makes community initiatives problematic, but that is not the subject of this chapter.)

In the case of groundwater, under Indian law, the ownership of land carries with it the ownership of the groundwater under it, subject to regulation and control by the state. It has been said that 'groundwater is attached, like a chattel, to land property', and that 'there is no limitation on how much groundwater a particular landowner may draw' (Chattrapati Singh 1991). It follows that only those owning land can have rights over groundwater; the landless (including communities, tribal and other, who may have been using certain natural resources for centuries) can have no such rights. Further, this legal position leads to inequities of various kinds: a rich farmer can install power-driven tubewells or borewells in his land and their operation can make dugwells in the neighbourhood run dry; he can sell water so extracted to his poorer neighbours even though the water may come from a common aquifer running under their land; and he can deplete the aquifer through excessive exploitation. The easement right makes regulation difficult.

Groundwater Legislation in Different States

Turning to actual groundwater legislation at the State level, efforts have been going on for a long time. In the 1970s the Central Government circulated a Model Bill to the State Governments for their consideration, and it was amended twice in later years. A few States took some action but in general the response on the part of the State Governments to the Central initiative was poor. After it came into being, the CGWA too has been urging the State Governments to pass legislation, set up State-level regulatory bodies, etc., but the results so far are unimpressive. In a number of States, drafts and bills are at various stages of consideration, but very few States have actual laws in operation. Some degree of regulation has been attempted under the various Irrigation Acts. The Gujarat Government has an Act in place for the regulation of groundwater, but it applies only to nine districts, and even that limited law seems to be moribund: there is a difference of opinion as to whether it is in fact in force. Maharashtra and Madhya Pradesh have Statewide Acts, but only for the regulation of drinking water sources. Tamil Nadu has an Act applicable only to the Madras Metropolitan Area; for the rest of the State, a Bill has been introduced. Andhra Pradesh has an Act of 1996, and an Ordinance of wider scope is under consideration for promulgation. Largely, then, we are still in the realm of intentions rather than actualities. There is no concerted, nationwide effort to treat groundwater as a scarce and precious natural resource to be protected and conserved.

Some Recommendations on the Subject

Inquiries as to the reasons for this unsatisfactory situation usually elicit references to political difficulties. Ignoring that paralysing answer, let us consider what needs to be done. There are four possible recommendations.

(i) The first is that the answer to all problems lie in water markets. 'Liberal' economists and officials of the multilateral financial institutions are the champions of water markets. (Implicit is a perception of water as a tradable commodity. This has already been discussed.) Water markets tend to emerge particularly in the context of groundwater extraction through tubewells and borewells, and though they serve some useful purposes, there are also dangers of unsustainable extraction as also of inequitable relationships between sellers and buyers. One is not ruling out water markets; they may have a role to play; but there are important issues of equity, social justice, resource-conservation and sustainability that cannot be left to market forces. The glib answer to that will be that these can be taken care of through 'regulation', but regulation is difficult and has not so far been seriously attempted.

(ii) The second recommendation is what one might roughly describe as the bureaucratic approach. The Model Bills circulated earlier by the Central Government and now by the CGWA provide essentially for old-style bureaucratic control, with not much room for the involvement of the people. Systems of bureaucratic control have not been markedly successful in other areas, and have tended to become dysfunctional; there is no basis for believing that this will not happen in relation to groundwater.

(iii) The third recommendation springs from the feeling that the existing legal position of vesting the ownership of groundwater in the landowner should be changed. Instead, it is argued that the state should hold the resource in trust for use by present and future generations. This implies that the resource so held in trust will be made available to people or institutions for use subject to certain regulations. This sounds reasonable, but one has serious doubts as to how this would work. We may use the word 'trust' instead of 'ownership' but in effect this would mean a dominant role for the state. Here again, the difficulties of 'eminent domain' will emerge. Moreover, considering the inequities and inefficiencies in the large state-managed irrigation projects, where the sheer difficulties of management have led to ideas of a transfer of management to the farmers, it seems retrograde and unrealistic to entrust groundwater resources to the state, even on a trusteeship basis.

(iv) The fourth possibility would be to treat groundwater as a common pool resource and place it under community management. This seems the most promising approach. Undoubtedly, there are major difficulties here too. Aquifers will have to delineated, user groups formed, rules of use and resource-management and-conservation formulated, regulatory authorities established, conflict-resolution mechanisms created, and so on; and the state will have a role to play in all this. This kind of approach is already in operation in the Joint Forest Management programme, as also in the participatory watershed development programmes governed by the Guidelines of 1994 issued by the Rural Development Ministry. The suggestion is to extend that approach with the necessary changes to the sphere of groundwater as well.

There is a variation on this in the World Bank's publication on 'Groundwater Regulation and Management' of 1999. It certainly argues for water markets, but it also envisages cooperation between the state and civil society. It recommends a cautious pilot project approach: the idea is that the approach of involvement of civil society be tried out on a pilot basis with the state standing by, ready to step in if the community initiative fails in a given case.

Postscript

It is necessary to take note of a book that came to hand after this book went to the press. *Tubewell Capitalism: Groundwater Development and Agrarian Change in Gujarat* by Navroz Dubash (Oxford, 2002) is an important book. In the two villages studied, the exploitation of ground-water (through progressively more advanced technologies culminating in deep tubewells) brought about dramatic increases in agricultural productivity and production, created prosperity (if unevenly), and led to the emergence of water markets and the gradual commercialization of the economies of the two villages. The picture that emerges from the case-studies is one of entrepreneurship and prosperity, though the prosperity is not unattended by problems.

That picture disturbs any simplistic positions one might have held on water markets or on the commoditization of water; however, one returns to one's doubts and concerns. (In fairness to the author of *Tubewell Capitalism* it must be mentioned that he is not an ardent advocate of water markets or an uncritical admirer of capitalism.) Going entirely by what the book says, inequities and inequalities appear to have been accentuated in some ways; the growing commercialization of life and relations seems to have benefited some but affected others (poorer people, small and marginal farmers) adversely; and there has been an alarming depletion of groundwater aquifers. Much damage has already been done, and it is by no means clear that further damage can be arrested. It appears that many farmers expect that the resource will be irretrievably run down and are preparing for the inevitable abandon-ment of agriculture in the not too distant future. Is this 'development' or disaster? Is the relentless pursuit of the declining groundwater best described as 'entrepreneurship' or as 'greed' in Gandhi's terminology? Has deep drilling technology been a blessing or a curse? Has it helped the people of these two villages to 'develop' groundwater or to destroy it? Are we looking at a healthy, harmonious relationship between hu-manity and nature or at a pathological, adversarial one?

The question may be asked whether, given the arid conditions and the presence of groundwater, it was not inevitable that the farmers should tap the resource for the practice of agriculture. Yes, indeed, but it is possible to visualize a very different way in which this might have been done. If the people in the two villages had refrained from going in for high technology or deep drilling, had limited their drafts on the resource to what was possible with 'lower' technology and kept recharging the aquifer through water-harvesting structures, would it not have resulted in 'development' of a more sustainable kind? Doubtless such development would have been more modest than what has in fact come about, but the latter is clearly unsustainable and self-destructive.

Finally, the central problem in groundwater management, from the points of view of both equity and sustainability, is the present legal position that vests the ownership of water in the owner of the land above it. That problem remains with us.

10

Towards a National Water Law?

Introductory

Before it finalized its report in September 1999, the National Commission for Integrated Water Resources Development Plan had held a series of regional conferences. At some of these meetings an idea that was put forward by several participants was that of a comprehensive national water law or code. This arose essentially from certain perceptions and concerns, such as a growing scarcity, ensuring access to safe drinking water, need for fair sharing and social justice, importance of economy and conservation, prevention of pollution, protection of water sources and supply systems, avoidance/resolution of conflicts, etc. There was a strong advocacy by some of a national law or code to take care of the proper management of this resource in the light of these concerns.

Many countries have national water laws, but in India we do not have one; instead, we have a national water policy. If Parliament were to wish to enact a national overarching or 'umbrella' water law, can it do so, given the quasi-federal constitution that we have? The answer is unclear. For the time being, let us assume that such legislation is possible, and consider what the components of a national water law or code should be.

If the National Water Policy (NWP) 2002 had been a carefully thought-through, structured, cogent and comprehensive document, it could have been transformed into a law; but given its weaknesses and deficiencies, a fresh effort at imagining what a national water law (NWL) would look like becomes necessary. What follows is essentially an exercise of broadly mapping out the territory to be covered, and taking note of the issues that would come up for consideration. Further, the exercise is a hypothetical one: a NWL may never come into being. (This chapter has links with Chapters 7 and 8 on Perceptions and Perspectives respectively; some overlap has been unavoidable.)

Nature of Water

In Chapters 7 and 8 the different perceptions of and perspectives on water have been set forth. An overarching water law or code cannot be based on a partial perception or perspective. A clear recognition of the multiple dimensions of water will have to be the starting point of the NWL.

Water as Essential for Life

The NWL will have to proceed next to the most important dimension, namely, the life-support function. Water is essential to life, and the NWL will have to provide an explicit legal basis for primacy to be accorded to this aspect of water.

While the fundamental right to life is held to include the right to drinking water, a separate and explicit recognition of the latter may be worth considering. We need of course to go beyond human beings and recognize the rights of other species, whether livestock or wildlife, to water. (It is often argued that we need to go still further and recognize the 'rights' of nature in general and of the river itself. The underlying concern here is a valid one, but the word 'right' is being used in an extended and special sense to 'valorize' certain concerns, to use a term that has come into vogue.)

Some would say that it is not enough to talk about a right to drinking water, and that the adjective 'safe' needs to be added. Potability certainly implies safety, i.e., conformity to certain standards. Further, an entitlement to drinking water also implies certain quantitative norms. The NWL may have to specify these norms, or alternatively, provide for their determination by separate laws or by subordinate legislation.

What does the recognition of a fundamental right to drinking water in the above sense mean in practical terms? Would it be the duty of the state and/or the community to provide it?[9] And by 'state' what level do we mean—federal, state, local? Or do the different levels have different responsibilities? How will this right be enforced? Can the state be regarded as having discharged its duty if it has licensed a private entity (corporate or other) to supply water to a particular area? In the event

---

9 See footnote 7.

of failure or deficiency on the part of the licensed entity, will the responsibility continue to vest in the state? What implications will the recognition of a fundamental right to drinking water have for the pricing of water? Will the principle of 'full cost recovery' apply, or will it need to be modified? Should water be subsidized for some and given free to those who cannot pay? There may be differences of views on these issues, and a number of different laws and policies may exist or may be under consideration, but a broad basis will have to be provided by the NWL.

Should Water be Free?

A brief digression on whether drinking water should be free may be in order here, as the question comes up repeatedly in discussions at seminars and conferences, and will need to be gone into when formulating an NWL. From the proposition that drinking water is essential for life and is therefore a basic right, it does not follow that it should be supplied free by the state. Food is also a basic right, but no one argues that the state should supply it free to all. (The subsidization of food is in fact a subject of considerable debate.) Even making the questionable assumption that the resource itself is free, water supply involves costs (purification, pumping, constructing and maintaining pipelines and so on), and these costs have to be recovered and revenues generated for maintenance and further investments. (In any case, nothing that the state supplies is really free: either there are specific user charges, or the costs are met through general taxes. In the case of water there is a strong case for proper user charges.)

The assumption that the resource itself is free is of course not correct: the 'development' of water resources involves financial, economic, environmental and human costs, and every decision in regard to its use involves opportunity costs in terms of alternative uses forgone.

Besides, water is a scarce resource, and is getting scarcer with the increase in population. A consciousness of the scarce and precious nature of this resource is hardly likely to be fostered by free supplies. Supplying water free is the surest way of encouraging the wasteful use of water. The availability of safe drinking water needs to be assured to all; supplies may have to be made at modest prices to the poorer sections of society and perhaps some free supplies may have to be made, but reasonable prices will have to be charged from these who are better off. (As for irrigation, it is an economic use of water, and does not fall within the ambit of the present discussion.)

It is also often stated that in Indian tradition we do not sell water. What does that proposition mean? Ignoring private hospitality, which is not what we are concerned with here, and ignoring also the practice, not merely Indian but universal (but now changing), of not billing for drinking water in restaurants and hotels (this is not really free supply, as the costs involved are doubtless taken into account in fixing prices for food), the proposition may mean the following:

- water for domestic uses is taken from the river without payment;
- when water so brought is given to the neighbour, a price is not charged;
- there is a tradition of putting up *piaos* by the roadside as an act of charity, for the supply of drinking water to thirsty passers-by;
- in some cases, a farmer may spare some irrigation water to another farmer without charging a price or treating it as a returnable loan. (This changes when serious costs are involved, as in the case of tubewells or borewells. Even canal waters, which are relatively cheaper, are not readily shared, and as and when they become costlier no farmer will spare them free of charge.)

These traditions do not warrant the inference that in public water-supply systems the state should refrain from charging for water.

(One is not too enthused about the rapidly expanding trade in bottled mineral water or about the advocacy of water markets, but the basic argument against them is not that water should be supplied free. The issues involved in water markets have already been discussed. As for bottled water or soft drinks, the market for them arises largely—though not wholly—from considerations of safety. If public water-supply systems could assure people of reliable, clean and safe supplies, the demand for bottled water would go down sharply.)

Unifying Framework

The next most important function of the NWL will be to bring together all existing water-related laws and case laws under an overarching and unifying framework. As we have seen, there are numerous laws relating to water, some laws not directly relating to water but having a bearing on it and a number of decisions by the Courts in various cases; these are not governed by any unifying framework or vision, nor are they necessarily mutually consistent. Further, there are unclear areas. Integrating the exist-ing diversity, remedying gaps and omissions, removing inconsistencies and contradictions and providing an overarching framework to the cluster (or clutter) will not be an easy task; but it needs to be done. In the process, many difficult issues will have to be dealt with, and choices and changes made. Some of these are briefly referred to in the following paragraphs.

Water Rights

An important issue is the one relating to 'rights'. The NWL will have to disentangle some complexities in this regard. (See also the discussion in the Addendum at the end of Chapter 8.) A clear distinction needs to be drawn between the basic right to drinking water (to which a reference has already been made) and other kinds of water rights. We often hear exhortations to 'define water rights and allow trading'. This cannot possibly apply to water regarded as life-support. The basic right to drinking water is a part of, or akin to, the right to life. The right to life is not a property right, and no one will want to, or can be allowed to, trade in it.

The other important uses that can be translated into the language of rights are irrigation and industrial use. What kinds of rights are these? Let us forget the existing laws and court decisions and look at this question afresh. (For convenience, we will confine ourselves first to the provision of water from public systems, whether canal systems or public tubewells, as also from community-owned systems.) The water needs of a farmer for irrigation purposes may be determined with reference to the area of land in question or the crops to be grown, or a combination of the two factors. In a situation of scarcity, limited quantities of water may be arbitrarily allocated by the state or some other agency (e.g., water users' association, if any) responsible for providing irrigation water. The important point is that the water is provided for irrigation. Similarly, an industrial unit gets an allocation of water from a state or municipal agency for certain uses (process, steam generation, cooling and so on). In both these cases, the entitlement to water is essentially linked to use. These 'water rights', if we wish to use that language, are use-rights. They may be customary rights, rights of long standing, contractual rights or statutory rights, but they are (to repeat the point) use-rights. How then can they be made into 'tradable property rights' as is often advocated? How can any individual or group or institution or corporate body be given a right to water unrelated to use?

If the use ceases, the entitlement should surely cease. Temporarily, a possibility of trading in water may arise. A farmer may for certain reasons decide not to cultivate his or her land for a year or two, or may temporarily switch to crops that need less water. He or she may then have surplus water for sale during that period. An industrial house may decide to suspend operations for a certain period for various reasons, and may have water to spare. However, these are temporary situations. If the industry closes down for good, or if the farmer makes a long-term change to crops that need much less water, or decides to move out of agriculture altogether, should the old water entitlements still hold and be allowed to be traded in? The NWL will have to reflect a considered position on this.

Turning to cases where a farmer or an industrial unit buys water from a private source (for instance, a farmer or a corporate body having a tubewell or a borewell), the buyer has of course only a contractual right. But what about the sellers? What kind of right do they have? At present, landowners have the easement right to water under their land, and are able to exploit it or sell it. That is how water markets have emerged. If the easement right were taken away, and aquifers come to be regarded as akin to underground rivers or lakes, and drafts on them are regulated, the whole situation will change. Under those circumstances only use-rights will be recognized, and the extent to which a landowner (or a corporate body) can extract groundwater will be limited by law with reference to equity and resource-conservation. Water markets may still function, but only to a limited extent and under a regime of regulation. Such a situation may not be easy to bring about, and some may not regard it as being a desirable change. The point is that the NWL will have to embody a considered view on this issue.

(We need not concern ourselves with water for 'recreation'—swimming pools, water sports and so on—as there are no questions of rights in these uses. As for navigation, what is involved is the maintenance of a certain level of flows in the waterway to keep it navigable, and not a question of 'water rights'. Livelihood and commercial rights may, however, be involved.)

Role of the State

We have argued earlier that even the state does not *own* water as a natural resource. However, the state has certain functions to perform with regard to water. It has to legislate on water; protect water sources and systems;

promote resource-conservation; ensure fairness and social justice; regulate the use of water from diverse sources; where necessary, undertake the provision of the 'water infrastructure' (to use the language of the World Bank); prevent or resolve conflicts; oversee quality; enter into treaties or agreements with neighbouring countries over common river systems and so on. In order to enable it to do all this, the state's role in relation to water has to be defined by law. At present it claims eminent domain which comes close to ownership. It is the author's view that while enabling the state to perform its various roles in relation to water, the idea of eminent domain needs to be abandoned or substantially modified. Others may hold a different view. What is important is that a definitive position on these matters should be arrived at and reflected in the NWL.

Community Management

The Prime Minister's address to the National Water Resources Council (NWRC) on the occasion of the adoption of the NWP 2002 on 1 April 2002 gives a resounding call for the recognition of the community as the custodian of water resources as also for a national movement for rainwater harvesting. As already mentioned in an earlier chapter, these ideas do not find a reflection in the NWP 2002. The more important point is that they do not seem to be supported by the laws of the land as they stand. Efforts to revive the lost or moribund traditions of water management (dying wisdom) and new movements to mobilize the community for local water-harvesting may come into conflict with existing laws. There is much talk of the community management of common pool resources, but the legal basis for this is tenuous. The role of the community or 'civil society' with respect to water resources, and the relationship between the state and civil society, are matters on which a clear position will have to be laid down by the NWL.

(It is also necessary to be clear about what comes within the ambit of the term 'common pool resources'. Please see the remarks on the notion of 'commons' in the chapter on Perceptions, i.e., Chapter 7 in this section.)

Question of Privatization

There is currently much advocacy of the 'privatization' of water, and a provision in this regard has been included in the NWP 2002. If that is the future course of policy, the NWL will have to provide for it. However, what exactly does 'privatization' mean in this context? What is to be privatized? It could mean the entrustment of the water-supply function in a city, say Delhi, to a private agency, but that could be a case of contracting out of a service hitherto performed by the state or a municipal agency; it would be 'privatization' only in a limited sense. Similarly, the supply of water in bulk from state-owned canal systems to a water users' association for distribution to its members could also be regarded as a kind of privatization. Entrustment of certain functions to civil society institutions or to NGOs could be yet another type of privatization. However, privatization could also mean allowing a private entity to build dams and reservoirs on rivers, or to exploit surface water bodies (e.g., lakes) or underground aquifers for commercial purposes. This raises questions not merely of equity and sustainability but also of control over natural resources (as pointed out in Chapter 5, section I). Much careful thinking and discussion will be called for before provisions in this regard can be included in the proposed NWL.

One important question that will need consideration in this context is whether allowing the domestic private sector to exploit national natural resources, particularly water, will make it difficult to deny a similar right to foreign investors in terms of the World Trade Organization regime and the principle of 'national treatment' of foreign investors; and if so, whether there is a danger of our losing control over our own natural resources, as some argue, or whether this is an exaggerated fear, as others say.

That may sound too neutral and non-committal. The intention is merely to acknowledge the fact that there are two views on the subject. For his part, the author has serious apprehensions about the implications (for equity, social justice and environmental protection and sustainability) of the marketization of water and the induction of the corporate sector into water services and water resource development. The concerns expressed by Rajendra Singh and Vandana Shiva in this regard need urgent and careful consideration. In particular, their cautions about the erosion of community rights, about the disempowerment of the people, and about the danger of the loss of control over our natural resources—first to the domestic corporate sector and then to foreign and transnational corporations—have to be taken very seriously indeed.

Large WRD Projects

There is at present no special law governing water resource development projects, but a large number of laws have a bearing on them (for details, see Chapter 13, section III). Some more laws may be needed as discussed elsewhere in this book. (For instance, the environmental impact assessment function may have to be provided with a statutory basis; and if a national rehabilitation policy comes to be adopted, that too may have to be made into a law.) The NWL should provide an integrated legal framework for such projects.

Participatory Irrigation Management

The NWL may also have to take note of what is known as participatory irrigation management (PIM). That programme encompasses the formation and registration of water users' associations (WUAs), the relationship between the WUAs and the state agencies, the WUAs' contractual rights to water and the state's contractual responsibility to supply the stated quantity of water, the related penal provisions, the volumetric pricing of water so supplied in bulk to the WUAs, the WUAs rights to distribute that water to their members and to determine prices and charge for water supplied and other related matters. There may be separate Acts dealing with PIM and WUAs (as in Andhra Pradesh and Madhya Pradesh). The question is whether the NWL should provide a broad basis for such Statewise legislation.

Another question that should be considered is whether, and if so how, the idea of PIM can be brought into private sector projects. It can hardly be argued that the idea of 'people's participation' does not apply to private sector projects.

Water-Related Conflicts

A major section of the NWL will have to be devoted to the subject of water-related conflicts at various levels between uses, users, sectors, areas, States and countries. The much-debated inter-State river water disputes, which are covered by an existing statutory mechanism (see Chapter 2, section I), are only one part of this vast and complex subject. Conflicts can arise, for instance:

- between drinking water needs and the demands of agriculture;
- between agricultural and industrial demands for water;
- between rural and urban needs;
- between neighbouring farms or communities or villages or districts;
- between different groups or sections within an irrigation command (e.g., head-reach and tail-end farms);
- between those who are displaced or otherwise adversely affected by a project and those who benefit from it;
- between the latter and those who suffer the consequences of reduced flows in the river downstream of a dam.

(The reference to the adversely affected categories covers not only people but also livestock, wildlife, aquatic life and natural systems including estuarine areas.)

In the design and operation of a project, conflicts (seemingly technical but in fact conflicts of interests) can arise between the demands of irrigation and power-generation (both of which would require the water level in the reservoir to be kept high) on the one hand, and the considerations of flood management (which would call for space to be left in the reservoir for the accommodation of floods) or navigation or the maintenance of the downstream river regime (which may call for more releases downstream from the dam) on the other.

As we have seen, conflicts can also arise between community initiatives and the state's perception of its own role. (The interaction of WUAs or watershed committees on the one hand and *panchayati raj* institutions on the other could also give rise to conflicts; this will be a subset of the conflict between civil society and the state.) And of course there can be conflicts between the 'development objectives' of the state and the people who suffer the 'social costs', the Narmada and Tehri cases being well-known examples.

Finally, there can be conflicts between countries on the sharing of rivers or aquifers and on water-quality issues.

The above was a sweeping and sketchy survey of conflicts of very different kinds, but with some relation to water. The putative NWL should formulate some general principles and lay down a broad framework for the avoidance or minimization of conflicts and for their resolution when they arise.

Protection of Water Sources

The NWL (in conjunction with the Water Pollution Control Act, the Environment Protection Act and the Forest Conservation Act) will also have to deal with the protection of diverse water sources (rivers, lakes, other surface water bodies, springs, mountains and forests, groundwater aquifers, water harvesting structures, wetlands and so on) from pollution/contamination/degradation as well as from over-exploitation. The extent of draft on nature will have to be regulated. The conservation of the resource, the promotion of a consciousness of scarcity, the need for enforcing economy in every kind of water-use, the avoidance of waste and misuse and so on are important concerns to be included in the law.

Other Matters

In regard to industrial uses, the law will have to provide a system for quantitative approvals in accordance with certain norms (case-by-case and cumulative), inclusive of a requirement of recycling and reuse, the prescription of effluent treatment and final standards for discharge and so on. These (as also contamination by agricultural residues) may be dealt with in other laws, but the NWL should also take note of them.

In-stream uses such as navigation, fisheries, boat-plying, run-of-the-river schemes; the requirements of aquatic life, wildlife and riparian communities; estuary maintenance, and the difficult concept of a 'minimum flow'[10] for maintaining the river regime may have to provided with a legal backing in the NWL.

On participatory watershed development there are of course the 1994 Guidelines, but they are executive orders; it may be desirable to give them a statutory basis.

As pointed out in an earlier chapter, there is a point of view that the NWRC and the NWP approved by it should be given a statutory backing. If this is considered necessary, the NWL may be the instrumentality for bringing this about.

---

10 On the problematic notion of a 'minimum' flow, see sub-paragraph (xiii) in the section on 'Changes in the Text' in the chapter on the National Water Policy 2002 (Chapter 5, section I).

Structured Whole

The diverse elements and themes mentioned above will have to be properly integrated into a structured whole keeping in mind a governing philosophy. (A suggestion in this regard has been put forward in Chapter 8).

Under What Provisions of the Constitution?

We return to the question of precisely how such a law or code can be enacted, given the fact that the basic entry relating to water is in the State List, and that Entry 56 in the Union List is only regarding inter-State rivers and river valleys. This is a serious but perhaps not an insuperable legal difficulty. In the past, ways and means have been found to enable Parliament to legislate on subjects that were not clearly within the Centre's domain: the Water (Prevention and Control of Pollution) Act and the Act establishing the Central Groundwater Authority are instances. If a consensus is reached on the need for a national water law and on the contents of such a law, legal experts can be consulted on the constitutional provisions under which it can be enacted.

III

Large Dams

11

The Large Dam Controversy

Dam-Building in India

India has a long history of water management. There were large numbers of widely varied water harvesting and conservation structures and systems in different parts of the country. While the rulers built some relatively large structures (for instance, the Grand Anicut built by the Chola kings in Tamil Nadu over a thousand years ago, or the canals built by the Mughals in north India), the systems were largely local and community-managed. All this changed with the advent of British rule and of 'modernity'. Control over water resources passed from the hands of the community into those of the state. While the ownership of natural resources was claimed by the state, their management passed into the hands of engineers and bureaucrats. The induction of Western engineering ushered in the era of large dams and there was a concomitant decline of traditional forms of small-scale, local, community-managed systems. The new projects became symbols of 'development' and came to be regarded as 'the temples of modern India' in Nehru's famous (but somewhat misinterpreted) phrase.

Some commonly accepted facts are that India has over four thousand 'large dams' as defined by The International Commission on Large Dams; that at the beginning of the 20th century India had 42 large dams; that by 1950 a further 250 had been added; that the rest came up in the second half of the last century, and that a large number of them, roughly half of the total number of large dams in the country, were undertaken in the period 1970 to 1989 (cf. The India Country Report to the World Commission on Dams, 'Large Dams: India's Experience', June 2000). Why were these dams built?

The answer is clear. The primary reason is that there are wide variations, both temporal and spatial, in the availability of water in the country. These variations led to proposals for the storing of river waters in reservoirs behind large dams: (*a*) for transferring water from the season of abundance to that of scarcity (as also from good years to bad), and (*b*) for long-distance water transfers from 'surplus' areas to water-short areas. In energy planning, large-scale hydropower generation, considered a necessary component of total generation, implies big projects. The objective of flood moderation has also led to some dam-building.

Contribution of Dams

What has been the contribution of large dams to the country? (The following draws upon—but differs in some respects from—the account given in the India Country Report to the World Commission on Dams (WCD).

The production of foodgrains increased from 51 million tonnes in 1950–51 to almost 200 million tonnes by 1996–97. The increase was the result of a combination of several factors such as high-yielding varieties of seeds, chemical fertilizers and pesticides, credit, agricultural extension, support prices and so on, but clearly irrigation played a crucial role, and some of that irrigation came from large dams while the rest came from other sources ('minor' surface-water irrigation and ground-water). On the question of how much of the increase in food production can be attributed exclusively to dams, there are different estimates ranging from 10 per cent to 30 per cent.

As for hydroelectric power, about two-thirds of the installed hydro-power capacity of 21,891 megawatt (MW) in March 1998 (out of a total generating capacity of 89,000 MW) is attributed to dams, with one-third coming from run-of-the-river schemes.

Turning to flood control, the contribution of dams has been modest. (Dams are not often planned with flood-moderation as a primary aim, and even where they are, the competing claims of irrigation and power generation often override the flood-moderation function. Further, while dams may indeed moderate flood flows to some extent under normal conditions, they may aggravate the position if, in the absence of a flood cushion, water has to be suddenly released in the interest of the safety of structures.)

Public water-supply is not often a stated objective of large-dam projects, but in many cases reservoirs and canals are in fact made use of for this purpose. There are a few projects which meet industrial demands for water. As regards navigation, this has not so far played a significant role in the planning of dam projects, except in the case of the Damodar Valley Corporation, and even there it did not develop as originally envisaged.

Thus it can be said that dams have contributed (along with other factors) to an increase in food production, added to hydropower capacity, provided water for domestic, municipal and industrial uses and (to some extent) helped in flood-moderation.

Growing Disenchantment

However, disenchantment with large projects has been growing over the past two decades. The Silent Valley Project in Kerala was abandoned. The Narmada (Sardar Sarovar) Project in Gujarat and the Tehri Hydro-Electric Project in the Himalayan region have been facing strong anti-project movements. Though these movements seem to have lost the battle to the dam-builders (at any rate for the time being) they have permanently altered our perceptions about such projects in many ways. Projects proposed long ago in North-east India (Dihang, Subansiri, Tipaimukh) and in Bhutan (Manas, Sankosh) have made hardly any headway because of opposition on diverse grounds. The earlier tacit consensus on such projects seems to have broken down; the statement that they are 'the temples of modern India' no longer commands universal assent. How did this happen? The answer lies in a convergence of dissatisfactions with such projects from diverse points of view:

(i) *Financial/economic*: ('time and cost overruns'; an insatiable demand for resources; the failure of many projects to achieve the projected benefits; their inability to generate revenues for reinvestment or even for proper maintenance, partly because of the poor pricing of irrigation water).

(ii) *'Political economy' aspects*: (the widespread perception of the prevalence of corruption and of the influence of vested interests in the planning and implementation of projects; serious inequities in the incidence of costs and benefits).

(iii) *Environmental/ecological concerns*.

(iv) *Concern about the displacement of people and dissatisfaction with rehabilitation policies and practices*; and so on.

All these strands are important, but the environmental and displacement aspects are at the heart of the controversy.

Environmental Impacts

The environmental and other 'impacts' are project-specific and vary from case to case, but the following is a generalized account in compendious form.

(i) The very processes of project construction in remote and often pristine areas (cutting, blasting, movement of large trucks and heavy earth-moving equipment, generation of noise and dust on a large scale, establishment of construction colonies, induction of large numbers of people from outside, etc.,) involve a violent disturbance of nature and a tremendous upheaval in the lives of local inhabitants (often tribal communities).

(ii) The creation of a large reservoir (and the construction of a system of canals) means the submergence of land (agricultural or forest land, and sometimes rural and urban settlements), the displacement of people and their livestock, the loss of occupations and so on. (Land is also taken away for project 'colonies').

(iii) The stilling of flowing water brings about drastic changes in its morphology and quality (temperature stratification, variations in nutrient content and dissolved oxygen at different levels, etc.,) which have grave consequences for aquatic and riparian life. The decay of submerged organic matter could also lead to emissions of greenhouse gases.

(iv) The most serious impact of the damming of a river is on the fish population, which is doomed to decline rapidly because movement is impeded and spawning hindered.

(v) The reservoir could spell some danger for wildlife; even if physical danger is averted, habitats and routes of movement are likely to be disrupted; groups and interdependent species could be split and food chains broken; some species could disappear, and this in turn could affect other species. (Communication links between human settlements could also be disrupted by the reservoir).

(vi) Flora too could be affected through the construction processes, submergence and other factors. Some species (endemic and/or rare) could be endangered, and herbs and medicinal plants of local or wider importance lost.

(Taking the loss of forests and the impacts on flora and fauna together, there could be considerable loss of biodiversity).

(vii) The reservoir and the canals could facilitate the spread of disease vectors and lead to an increased incidence of malaria, filariasis, schistosomiasis, etc., (though the last mentioned disease is not a serious Indian problem).

(viii) The creation of a large waterbody could bring about climatological changes.

(ix) Projects in seismically active areas such as the Himalayan region are subject to the risk of earthquakes and their possible impact on the dam and the slopes of the reservoir. There may also be possibilities of re-activation of old and dormant faults. An important issue on which there is considerable difference of opinion is that of 'reservoir-induced seismicity'.

(x) The damming of a river affects the whole river regime. Flows as well as the silt load and nutrient content downstream of the dam would be substantially reduced, and this would have an impact on lives, occupations and livelihoods downstream (fisheries, the plying of boats, agriculture, settlements alongside of the river, industries, etc.). Estuarine conditions may also be adversely affected (decline in fish population, the incursion of salinity from the sea, etc.). The reduction of flows also means a deterioration in water quality and an increased concentration of pollutants in the river downstream of the dam, and may affect the self-regenerating capacity of the river. A further consequence of reduced flows is a decline in ground-water recharge.

(xi) In some cases, structures of religious, historical or cultural importance may be in danger of submergence or damage.

(xii) There is a certain inherent and unavoidable risk in the damming of a river. Under normal circumstances, a dam could moderate floods; but under exceptional circumstances, the dam itself could become the source of danger to downstream areas either because of a dam-break or because of emergent releases of flood waters in the interest of the safety of the structure.

(xiii) The construction of canals could (unless great care is taken) disrupt the natural drainage leading to drainage congestion. In the command area, the practice of canal irrigation for some years could result in the emergence of waterlogging conditions and salinization of land, and in some instances valuable agricultural land may go out of use.

(xiv) Finally (though this is not a necessary consequence of dams), the belief in the virtues of dam-building and extension of irrigation often leads to the application of these ideas in the wrong places. (For instance, irrigated agriculture may not be the best option for a desert area; and attempts to convert nomads to settled agriculture, or failing that, to induct agriculturists from other areas, as was done in Rajasthan, may not be the wisest thing to do).

The above was a summary and simplified account of the kinds of consequences that large-dam projects could have. Not all of them occur in all cases, nor are they all of equal importance, but quite a number are common to many projects and many of them are indeed matters for concern.

Some of these effects cannot be remedied or even mitigated, and in some cases, efforts at the mitigation of or compensation for environmental impacts may in turn create further problems. Further, it is clear from past experience that all the consequences and ramifications arising from the damming of a river cannot really be fully foreseen and planned for.

Human Impacts

In most cases, there will also be varying degrees of displacement of human settlements, with the attendant problems of resettlement and rehabilitation; this impact often falls on poor and disadvantaged sections, particularly tribal communities. There are inherent difficulties in resettlement and rehabilitation: a lack of full knowledge of the numbers and categories of people likely to be affected; separation of communities from the natural resource base on which they are dependent; inadequacy of land for land-based rehabilitation; scattering of well-knit communities; resettlement in distant and unfamiliar areas; difficulties with the host communities in the resettlement areas; major transformation in ways of living, loss of old coping capabilities and the need to learn new skills and ways of living, and so on. However good and enlightened the rehabilitation policies and 'packages' may be, there will inevitably be great hardship and suffering, to which the response of the governmental machinery is rarely adequate, much less imaginative.

Usual Answers: Environmental Impact Assessment/ Remedial Measures

These implications and ramifications of large-dam projects were not well understood in earlier years, but they are now widely recognized. Not even the most ardent advocates of large dams will deny their negative aspects. Faced with a catalogue of ills, their response will usually be threefold.

First, they will readily concede that all these 'impacts' need to be studied fully and thoroughly. There is general agreement that proper 'environmental impact assessment' (EIA) studies must be made in all cases; that this should be part of project-formulation *ab initio* and not an exercise to be undertaken to meet an external requirement after the project has been prepared, and that the EIA study should be an important element in the process of project appraisal. EIA is now a standard requirement, and both the Central Water Commission (CWC) and the Ministry of Environment and Forests have laid down detailed guidelines on scope, coverage and methodology. A clearance by an Environmental Appraisal Committee under the Ministry of Environment and Forests is a pre-requisite for final investment approval to all big projects.

Secondly, it will be argued that once a full EIA is available, what is needed is merely the reckoning of all the environmental and displacement aspects as 'costs' (in addition to the direct financial costs) and the balancing of these costs against the 'benefits' which the project will bring (increased agricultural production resulting from irrigation, increased industrial activity made possible by hydroelectric power, their multiplier effects, etc.,) in a thoroughgoing cost-benefit analysis.

Thirdly, it will be pointed out that the planning of a project would include the formulation of detailed measures to *remedy* or *mitigate* or *compensate for* the adverse impacts of the project.

Difficulties with Those Answers

Unfortunately that line of argument ignores several serious difficulties.

In the first place, environmental and other concerns continue to be regarded as disagreeable external impositions; they have not become integral parts of project planning from the start, despite many 'guidelines' and instructions to this effect. Everyone pays lip service to those concerns, but the prime interest is in the engineering aspects. The implicit assumption is that water planning is essentially a matter for engineers. (It is significant that the CWC, which regards itself as the apex body for water planning in this country, is not a multi-disciplinary body encompassing agriculture, environmental sciences, economics, sociology, law, etc., but merely a body of engineers).

Secondly, EIAs are notoriously undependable. When they are undertaken in-house by the project planners, the desire to get the project approved may influence the EIA and render it suspect. Even when a reputed external consultancy firm is engaged (as is often the practice), the thoroughness and objectivity of the study cannot be taken for granted. It needs to be recognized that the insidious pressure on the consultant to be 'positive' about the project could be very strong: to say this is not to imply that there is always a collusion between the project planner and the consultant. The latter has an interest (not necessarily conscious) in coming to the conclusion that the adverse impacts of the project can be remedied or mitigated or compensated for; that the project will still remain viable, and that the overall balance of costs and benefits will be favourable to the project. A consultant who says: 'The impacts of this project are too grave to be mitigated or offset: the project should not be undertaken' is unlikely to secure many assignments. It is only a disinterested examination by an independent appraisal agency, say, the Ministry of Environment and Forests or an agency appointed by it, that could be expected to be truly neutral and objective. However, even that agency could come under strong pressure from other agencies *within* the Government to be 'positive' and supportive of 'development'.

Thirdly, it is unrealistic to imagine that any EIA, however careful, can be made truly comprehensive and exhaustive. Large-dam projects are often horrendous interventions in natural environments, and it is impossible to foresee all the consequences of such enterprises. Despite extensive studies, there may be many aspects, dimensions and ramifications which have not been taken note of. (This is not a general *a priori* statement, but has in fact been found to be the case in several instances).

Fourthly, it is not always possible to remedy or mitigate or compensate for the ill effects of such projects. For instance, what goes by the name of 'compensatory afforestation' is a delusion. It is rarely feasible to create a new 'replacement forest' in the neighbourhood of the existing one which will be submerged, or in the same ecological zone; quite often the compensatory afforestation takes place in a distant and very different area. Further, while such afforestation may be successful and may evolve into a new ecological system in due course, what is lost cannot be replaced: *that* ecological system is gone forever. Again, the changes in river morphology and water quality brought about by stilling a flowing stream, and the impact of such changes on aquatic and riparian life, simply cannot be remedied. The decimation of fish populations by the damming of a river is also totally inescapable; fish ladders, etc., rarely work satisfactorily, and the development of new reservoir fisheries is no answer to the distress and disaster inflicted on existing fish populations. Similarly, once a dam is built, the river will never be the same again; flows downstream will necessarily be reduced, with unavoidable consequences for aquatic life and riparian communities. Displacement because of submergence, again, is inescapable, and rehabilitation 'packages', however enlightened and generous, do not always work well in practice.

Fifthly, the cost-benefit calculus is a flawed basis for decision-making because: (i) it is susceptible to manipulation (costs are usually understated and benefits overstated); (ii) it is necessarily incomplete and inadequate (not every aspect or dimension can be brought within the ambit of the calculus); and (iii) it is morally blind (the infliction of misery on some people is often sought to be justified on the ground that a larger number elsewhere will be benefited).

(Pious declarations about giving the project-affected persons a share in the benefits downstream are rarely translated into practice. References to 'stakeholder participation' have now become fashionable, but it is doubtful if this indicates any depth of concern about such matters. Indeed, the very term 'stakeholder' is ironic; in what sense can the hapless communities uprooted from their centuries-old habitat for the construction of projects be regarded as 'stakeholders' in those projects, when they are in fact the *victims* of the projects?)

Finally, it is the benefits (direct and indirect) of a project which are held to justify the costs (financial and social); and that justification (i.e., the case for the infliction of misery on people and damage on the environment) tends to get undermined by the fact that the costs are certain to be incurred and are almost always higher than projected, whereas the claimed benefits are often problematic and may not be fully realized.

(In regard to the 'political economy' aspects of such projects, it could be argued that inequities, injustices, corruption, collusion, etc., arise from the socio-political milieu and cannot be attributed to dams; but some of the inequities and ills are perhaps facilitated by, or at any rate associated with, large-dam projects.)

Are Big Dams Avoidable?

Keeping in mind the various impacts and consequences of large dams, the crucial question is whether, given the projected magnitudes of demand for water (linked to rates of growth of population and urbanization), such projects are avoidable. A widely held view is that they are not. This way of thinking holds that future needs cannot be met without massive 'water resource development', that expression being treated as synonymous with large 'storage' (i.e., dam-and-reservoir) projects; and that local rainwater harvesting and watershed development schemes, while very necessary, are bound to remain secondary and supplementary to large projects and cannot be a major component of water resource planning. However, there are others who see great potential in water harvesting and watershed development and are convinced that these activities, undertaken in several thousands of locations all over the country, are capable of making a substantial contribution towards future needs, while being at the same time environmentally benign, people-centred and conducive to equity. That potential needs to be carefully assessed, but it is clear enough that a major push needs to be given to these activities; there is no justification for assuming *a priori* that they can play only a small, supplementary role in water resource planning. If in fact they can make a significant contribution, the need for large projects can be minimized; only a small number may then be needed, and the environmental, social and human impacts will be correspondingly reduced.

In this context it is necessary to take note of some apparently clinching arguments in favour of large projects.

A point made by some supporters of such projects is: yes, doing things has a cost; but there is also 'the cost of not doing'. This argument is often reinforced by the rhetorical question: where would the country have been without Bhakra–Nangal? Many find this line of argument persuasive. However, this is not a new or additional argument, but only a familiar one in a different form. 'The cost of not doing' means merely that in the absence of the project, certain benefits would not be available. This is nothing more than the old argument that the benefits justify the costs; we have already dealt with this. Further, it is fallacious to equate the non-undertaking of a large project with 'not doing'. The choice is not between 'doing a project' and 'not doing anything'; there are other things (such as demand management, conservation, local water-harvesting, etc.) that can be done. As for the question of what we would have done without Bhakra–Nangal, it is a hypothetical one to which only a speculative answer can be given. We know the Bhakra–Nangal 'scenario' because that is what actually happened; we do not know what the alternative history would have been if it had not come into existence. However, we need not readily assume that there would have been an absence of development on the agricultural front. Understandably, data and information are available only in respect of the route (of large projects) actually taken, and not in respect of the alternative routes that have not been explored. All that one can do is to point to the successful instances of watershed development and social transformation, and say that there is no reason why these cannot be replicated in large numbers.

Another seemingly powerful argument is that even if there is no need for large projects for irrigation, they are definitely needed for the generation of hydroelectric power; that given the magnitudes of demand projections, large additions to generating capacity are called for; that a suitable thermal-hydro mix is required for maintaining a proper balance between base-load and peaking capacities; and that hydroelectric power is 'clean', i.e., it does not create the kind of pollution that is incidental to coal-burning. Certainly, both the power shortage problem and the peak-demand problem need to be dealt with, but centralized generation in large projects is not the only answer to those problems. It has been argued (by Dr A.K.N. Reddy and Girish Sant,[11] among others) that through a combination of demand management, energy-saving, technological improvements and getting more generation out of capacities already installed, the need for additions to capacity can be greatly reduced; that significant additions can be made through extensive decentralized generation, and that if this approach were adopted very few large projects would be needed. This proposition, which runs counter to the Establishment view, has not been given serious consideration. (As for hydropower being 'clean', the fallaciousness of that argument has been definitively brought out in Patrick McCully's *Silenced Rivers*. While the

---

11 Dr A.K.N. Reddy and Girish Sant, and K.R. Datye: 'Submissions to the Five Member Group on the Sardar Sarovar Project', report of the FMG, Vol. II, Appendices, Ministry of Water Resources, 1994.

operation of a hydroelectric station may not emit harmful gases or spread particulate matter as coal-burning does, the *construction* of the project and the existence of the reservoir itself have a whole range of severe environmental consequences, as we have already seen: such a project can hardly be described as 'clean'.)

It is also argued that the big cities in India (Delhi, Mumbai, Calcutta, Chennai) are very short of water and that only large projects can meet their needs. This is an unexamined assumption. There is enormous scope for the augmentation of supplies to such cities through local efforts (in addition, of course, to proper demand management). Realizing that the prospects of water from distant projects are remote and uncertain, the city of Delhi is now seriously exploring the possibilities of local augmentation through increased storage in existing channels such as the Najafgarh *Nalla*, re-activation of old and disused waterbodies such as the one at Hauz Khas, rooftop collection of rainwater and other similar means. Similar efforts are also being made elsewhere.

The proposition that the future needs for water, food and energy can be met through alternative means and that large dams are not required is confidently asserted by some, but it cannot be said to have been fully established; further work on this is necessary. Indeed, this is the crucial question for consideration. However, those who are against large dams would say that in any case dams do not serve the projected purposes and do far more harm than good, and that the establishment of alternatives cannot be a precondition for rejecting something we know to be bad. That is not an argument that can be lightly dismissed.

An Unresolved Issue

As mentioned earlier, several kinds of criticism converged into a powerful movement against such projects, both within the countries concerned and internationally. There was a sharp polarization of attitudes on this matter. Unfortunately the debate between the 'pro-dam' and 'anti-dam' camps was marred by hostility and prejudice on both sides. By the early 1990s there was an *impasse* on this issue.

The pressure of the anti-project movements, the sharing of some of their concerns by certain governments, the desire of both governments and the dam-building and equipment-supplying industries to bring the uncertainty surrounding such projects to an end, and the World Bank's own desire to obtain a clear mandate on this matter, led to an unprecedented consultation conference at Gland in Switzerland in April 1997. Following that consultation, the World Bank and the World Conservation Union (IUCN) together established a World Commission on Dams (WCD), with a composition reflecting different concerns and interests, funding from diverse sources and institutional arrangements to ensure the independence of the Commission. The WCD began functioning in 1998 and 'launched' its Report in London in November 2000. Unfortunately, this did not mark the end of the controversy; the divergent reactions to the Report in fact led to a further sharpening of the divide between 'pro-dam' and 'anti-dam' groups. (The Indian reactions to the WCD's Report are analysed in Chapter 15.) The controversy surrounding that Report is still continuing, and though efforts have been initiated to bring about a constructive debate, the present position is that the issue remains unresolved.

Conclusion

The question is not whether large 'water resource development' projects should be allowed or ruled out, but whether they should be the first choice or the last option. That question, if rigorously pursued, will take us beyond the domain of water resources and into larger issues. The examination must necessarily include a questioning of the demand projections (particularly in the case of energy) and of the lifestyles they are derived from, and a redefinition of what exactly constitutes 'development'.

Large dams are only one aspect or feature of the modern world. It is possible to marshal an impressive array of evidence against dams; but it is equally possible to build up a strong case against other symbols of 'development': coal-burning and nuclear power plants; metallurgical, chemical, hydrocarbon and petrochemical industries and mining complexes; monstrous megalopolises; the exploding number of automobiles; vast networks of railways and highways built by trenching into flood plains, drainage channels, fields, forests and wildlife habitats, and by blasting hillsides and tunnelling through mountains; the onslaught on aquatic life by giant trawlers and whaling vessels; the staggering global trade in oil and the ever-present threat of oil spills, and so on. All these are manifestations of a certain conception of 'development' and a related attitude to nature. At the moment it is difficult to see how a change of direction is going to be brought about and possible doom averted.

However, instead of perplexing ourselves with large and unanswerable questions, we could perhaps consider what can be done in practical terms and in limited contexts. From that point of view, and in the context of water resources, it seems very necessary to explore all non-dam possibilities of meeting future needs before we consider recourse to large-dam projects. (We shall revert to this in Chapter 25 of this book.)

12

Large Dams, Trans-Boundary Waters, Conflicts

It is interesting to note that conflicts over river waters, whether inter-country or intra-country, seem often to arise in the context of large projects. The India–Bangladesh dispute over Ganga waters was precipitated by the Farakka Barrage Project in India. Projects on the Kosi and Gandak rivers were the starting points of a prolonged history of mis-understandings between India and Nepal, and that mistrust was further accentuated by the Tanakpur Barrage Project, until it was resolved through the Mahakali Treaty of February 1996; but even after that Treaty there are pending issues relating to the Pancheswar Project. Within India, the Cauvery dispute arose because dams and reservoirs built by the State of Karnataka had the effect of reducing the flows into the Mettur Reservoir in Tamil Nadu. Between Karnataka and Andhra Pradesh, there is a dispute over the Alamatti Project of the former on the Krishna river. Even within a State, a large project creates conflicts of interests between the people of the upper catchment area and those downstream; between those who bear the social costs of the project and those who enjoy the benefits such as irrigation or hydropower; between the head-reach farmers and those at the tail-end of the canal system, as also between rich farmers and poorer ones in the command area; between human beings and wildlife; and so on. Such conflicts have been incidental to many projects, but have been particularly marked in the case of some. In India, the Sardar Sarovar Project on the Narmada in Gujarat State and the Tehri Hydroelectric Project on two tributaries of the Ganga in the Himalayan region, have been the centres of fierce controversy, and have become international *causes célèbres*. In Nepal, the Arun III Project became controversial, leading to the withdrawal of the World Bank. There is also a controversy raging around the Kalabagh Project in Pakistan. It would appear that large projects tend to become the foci of conflicts. Dams will doubtless continue to be built, and some of them on trans-boundary rivers. This chapter (based on a small paper originally written by the author for the World Commission on Dams (WCD)) considers the issues that arise in such cases.

First, it is necessary to be clear regarding the manner in which the three components mentioned in the title of this chapter are related.

(i) There is a cluster of issues relating to *large dams*: violent disturbance of pristine areas and disruption of long-established ways of living; submergence of land and forests; impacts on flora and fauna and reduction of bio-diversity; reduced river flows downstream with serious consequences and implications; displacement of people and problems of resettlement and rehabilitation; public health aspects; the benefits expected from the project; the overall balance of costs and benefits; inequities in the incidence of costs and benefits, and in the distribution of the benefits; heavy drafts on budgetary resources, and the question whether the objectives in view could have been achieved through alternative means that avoid some or most of the adverse impacts mentioned above.

(ii) There is another cluster of issues relating to *trans-boundary waters* (including both boundaries between countries and political divisions within a country, particularly one with a federal or quasi-federal structure): riparian rights and obligations; principles of water-sharing, and the sharing of costs and benefits on projects; issues of water quality; the impacts of the actions or omissions of one riparian on other riparians; the question of 'harm' or 'injury' to fellow riparians; and so on.

(iii) There is yet another cluster of issues relating to *conflicts* (whether water-related or other) between states or between political units within a state and the modalities of their resolution: the *avoidance* of conflicts; early identification of possibilities of conflict, and the processes and mechanisms of resolution of conflicts.

These three classes of issues are distinct, but as many dams have been built on rivers that cross political boundaries (or flow along or mark such boundaries), the first two clusters get interconnected; and as both large dams and trans-boundary waters hold potentials for conflicts, the third cluster also comes in. Diagrammatically this can be represented as in Figure 12.1.

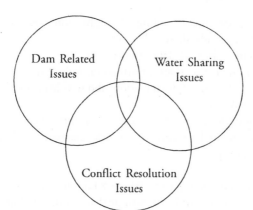

Figure 12.1: Inter-relationship between Large Dams, Trans-boundary Waters and Conflicts

One kind of conflict may be eliminated from the discussion at this juncture, namely, deliberate intention to harm. The wilful obstruction of the flow of water to a downstream area; the deliberate creation of floods downstream or backwater effects upstream; the poisoning of water sources; the bombing of a dam to cause devastation: these rather fanciful hypotheses illustrate acts of war or hostility, and are not the kinds of conflicts that this chapter is concerned with. Assuming that there is no deliberate intention to harm, how do conflicts in relation to water arise, whether between countries or between constituent units within a country? They arise from one or more of the following causes:

- wrong principles (Harmon Doctrine, prior appropriation, prescriptive rights, etc., asserted in an absolute manner);
- limited vision (myopic nationalism, blind assertion of local perceptions);
- lack of sensitivity on the part of the stronger party, excessive touchiness on the part of the weaker;
- wrong approach to natural resource planning (the notion of 'conquest of nature' or 'harnessing' natural resources through the application of science and technology, arising from the legacy of Prometheanism and implying an exploitative or adversarial relationship to nature; and a strong inclination towards gigantism);

- inadequate understanding of implications and consequences; failure to study these fully;
- ignorance; lack of data/information;
- unwillingness to share information; failure to consult all concerned; failure of imagination about others' needs, rights or concerns; and
- politicization, i.e., the tendency for differences over water or environmental concerns to become elements in domestic electoral politics.

How can conflicts relating to river waters (and in particular large dams) be avoided or minimized? Let us assume that as a first preference all non-dam possibilities will be explored and that big projects on trans-boundary rivers will not be undertaken unless they are found to be unavoidable. That still leaves us with the question: how can conflicts be minimized?

In this context we often come across two kinds of 'ideal' recommendations. From a *hydrological* point of view, the recommendation is that national or political boundaries should be ignored, that a hydrological unit such as a basin or sub-basin should be taken as a whole and that there should be *integrated* water resource planning for such a unit. From a *political/economic* perspective, the recommendation is that *regional* planning is superior to national planning. There is much force in these propositions, but there are also some difficulties that must be taken note of.

The advocacy of regionalism tends to become doctrinaire. There are some problems and issues that are best dealt with on a national or local basis; some that call for cooperation between two countries or units, and others that demand a regional approach. The circumstances vary from case to case, and in each case the most appropriate route needs to be followed. On the one hand, a rigid bilateralism such as that adopted by the Government of India is unwise and unduly self-limiting; on the other, a dogmatic advocacy of regionalism as inherently superior to a national or bilateral approach would unnecessarily complicate simple issues and render them more difficult to resolve. What is called for is pragmatism rather than doctrine.

The 'basin' approach is theoretically sound, but some basins are too large and have to be broken down into sub-basins. Besides, there is a great deal that can and needs to be done on a *local* basis through community initiatives. Such initiatives are not likely to be fostered by a planning approach that thinks in terms of large areas. As pointed out in an earlier chapter, the ideas of 'basin planning' and of 'integration' carry with them an implicit bias towards gigantism and towards a technology-driven rather than a people-centred approach. It is necessary to ensure that 'basin planning' covers the whole range of activities from local water-harvesting and watershed development to large projects; and that 'integration' covers land-use and water-use, brings together all the disciplines concerned and is not dominated by engineering, incorporates environmental, social and human concerns (including special concerns relating to women, children, and tribal, backward and other disadvantaged groups), and proceeds on a people-centred, 'bottom-up' basis rather than a bureaucracy/technocracy-driven 'top-down' basis. 'Holistic' would be a better word to use than 'integrated'.

Besides, a 'hydrological' approach that ignores political boundaries may sound right in theory but may not work in practice. In principle it may be possible to argue that benefit-sharing is better than water-sharing, and that within a basin or a sub-basin as a whole, food production should be concentrated in one area, power generation in another and industry in yet another; but this may not be acceptable. Political boundaries, whether between countries or within a country, exist and cannot be done away with. A theoretically right approach may have to be moderated by a degree of realism. The best need not be the enemy of the good. 'Integration' may be ideal, but it may sometimes be necessary to settle for the second best option of 'coordination'. Similarly, enlightened nationalism or bilateralism may be the first step towards eventual regionalism.

Turning to *principles*, neither the Harmon Doctrine (that of territorial sovereignty) nor that of prescriptive rights or prior appropriation has found general approval. What commands a fair degree of international acceptance is the principle of *equitable apportionment for beneficial uses*. That was the language of the old Helsinki Rules. The present UN Convention on the Non-Navigational Uses of International Watercourses (passed by the UN General Assembly in 1997, but still awaiting ratification by the stipulated number of countries) requires the watercourse States to 'utilize an international watercourse in an equitable and reasonable manner' (article 5, clause 1). Again, the next clause requires the watercourse States to 'participate in the use, development and protection of an international watercourse in an equitable and reasonable manner'. What exactly is 'equitable' has of course to be determined with reference to many criteria, and there is enormous scope for differences here, but there is some merit in a general subscription to the principles of equity and reasonableness. Similarly, a general admonition to the upper riparian on the question of causing harm to the lower riparian is unexceptionable, though the wording has changed from 'substantial harm' in the Helsinki Rules to 'significant' adverse effects in the UN Convention.

The short point is that if countries falling within a river basin or sub-basin wish to avoid conflict while planning a project, there are enough principles and guidelines to go by. To put it in a nutshell (even if this runs the risk of oversimplification), the upper riparian, in exercising its powers of control over waters, cannot ignore the *rights* of the lower riparian; and the lower riparian, in asserting its rights over the waters, cannot ignore the *needs* of the upper riparian. Given that kind of understanding, conflicts will either not arise at all or can be resolved without much difficulty when they do.

However, principles are not enough. *Knowledge* and *awareness* are important. Large dams are major interventions in nature and should not be undertaken without the fullest study of the likely consequences and implications of such intervention. This imperative is clear enough in all cases, but becomes even more so when the project is on a trans-boundary river. Environmental impact assessments (EIAs) should not stop at boundaries. What happens beyond those boundaries is equally important.

Knowledge so generated must be *shared*. Any planning of dam building should start with *advance notice* of that intention to all concerned. Here again, the prescription is valid for all cases: the people concerned (those who are likely to be affected adversely as well as those who stand to benefit) must be taken into confidence at the earliest possible stage; and this is even more important if different countries (or political units within a country) are involved.

Advance information should be followed by *continuous consultation*. It is the failure to give notice or consult that has led to (or aggravated) most conflicts. 'Consultation' does not mean presenting the other countries concerned (or the people concerned within the country) with a prepared document and asking for comments: the planning should be a 'participatory' exercise *ab initio*. This holds good both within a country and between countries. Ideally, there should be *joint* planning; as a second best course, there should be at least *coordinated* and cooperative planning.

The consultation and cooperation should not be merely an inter-governmental matter; the *people* need to be involved. There has been much talk of 'stakeholder participation' in recent years. This is part of the Dublin–Rio principles. It needs to be recognized that stakeholders exist not merely within the borders of the project-planning country but also beyond those borders. The principle of participation applies to them as well.

Similarly, when conflicts do arise or seem likely to arise, the processes of avoidance and resolution cannot be left entirely to governments. Civil society on both sides, represented by respected persons of goodwill and/or by reputable NGOs, has an important role to play. Academic and research institutions too can help. It is being increasingly recognized that such 'Track II' efforts (as they have come to be known), can make valuable contributions to the processes of conflict-resolution: they can help to break logjams at the inter-governmental level, gently persuade governments to go to the negotiating table, facilitate talks through behind-the-scene 'good offices', provide ideas and proposals (menus) to serve as the basis or starting point for purposeful negotiations and assist in the finding of answers to the difficulties that arise in the course of such negotiations. This is not a general hypothetical statement; it is based on actual experience. 'Track II' activities cannot stop with the signing of a treaty or agreement; they will continue to be needed during the processes of implementation.

Finally, any such treaty or agreement must of course include suitable provisions—consultations, conciliation, mediation, arbitration, adjudication, as may be agreed upon—for the resolution of differences and disputes. As regards institutional mechanisms, there are many models to choose from: bilateral or multilateral commissions; purely govern-mental bodies or bodies with a large non-official component; advisory or empowered bodies, and so on. What is feasible in a given case will be a function of the felt needs and the facts of geography on the one hand, and the state of political relations between the countries concerned on the other. What is important is that institutional mechanisms appropriate to a given case must be established at a very early stage.

13

Large-Dam Projects: The Framework of Laws, Policies, Institutions and Procedures

Preliminary

This chapter will provide a broad but compendious survey of the legal and institutional structure, setting and ambience within which large-dam projects in general have come into being and functioned in India. While the chapter has been divided into sub-sections on laws, policy, institutions, etc., for the sake of convenience, there will inevitably be some overlaps among the sections.

Laws

There is no separate law or set of laws relating specifically to dam projects, but the planning, approval, financing, construction, operation and maintenance of such projects take place within the constitutional and legal framework of the country, and in particular, the provisions relating to water.

Attention is invited to the chapter regarding the constitutional entries relating to water (Chapter 1, section I). That ground will not be covered here afresh. As discussed in Chapter 1, the primary entry relating to water is Entry 17 in the State List:

'Water, that is to say, water supplies, irrigation and canals, drainage and embankments, water storage and water power subject to the provisions of Entry 56 of List I.'

The legislative competence of the States in relation to water, conferred by Entry 17 in the State List, also implies executive power, and it is this that enables State Governments to plan and implement dam projects.

Again, as we saw in Chapter 1, the Centre has not made (or been able to make) significant use of the enabling provisions of Entry 56.

The Inter-State Water Disputes Act 1956 (ISWD Act) passed by the Union Parliament under the provisions of Article 262 of the Constitution is an important law in the context of the planning and construction of dams. We are not concerned here with the problems experienced in the operation of the adjudication process or the changes that are felt to be needed (those subjects have been discussed in Chapter 2, section I), but with the bearing that the ISWD Act has on dam projects. Not only must the allocation of waters and the restrictions (if any) imposed by a Tribunal's Award be kept in mind, but in some cases the Awards have an even more direct relevance for project planning: projects figure in some Tribunals' Reports (e.g., Narmada, Krishna); the Narmada Tribunal's Award specifies certain project features; in some cases the Award itself leads to the formulation or acceleration of projects with a view to making sure of retaining the State's share of the waters (e.g., planning for Krishna waters by Maharashtra), and there are also new post-award disputes relating to certain projects (e.g., the Karnataka/Maharashtra–Andhra Pradesh dispute over the Telugu Ganga Project, and the dispute regarding the Alamatti Project between Andhra Pradesh and Karnataka).

The role given to the Centre in relation to inter-State rivers gets reinforced by the use of the provisions of Entry 20 in the Concurrent List, namely, economic and social planning. It is that entry which provides the necessary constitutional basis for the requirement of a Central clearance for major and medium irrigation projects (including those not involving inter-State rivers) for inclusion in the national Plan. This has been questioned by some State Governments but the clearance requirement continues to be operative, though some relaxations have been made in the case of 'medium' projects which are now subjected only to a summary or *pro forma* check.

(In passing, it may be noted that the sanction behind the clearance requirement is the leverage that the Centre has through the provision of Central financial assistance to State Plans. This is a weak sanction because the Central Plan assistance to States, calculated under what is known as 'the Gadgil Formula' which came into operation from the Third Plan, is with a few exceptions not project-linked. Prior to the introduction of this system there was in fact a close project-linkage, but that has not been the case from the Third Plan onwards. Under the circumstances, the Centre's ability to influence the planning and implementation of State projects has been rather limited. In recent years there has been a change in the situation. There are some special instances of project-linked assistance: for instance, externally aided projects which receive an 'additionality' of Central assistance, certain projects considered to be of special national importance, the Accelerated Irrigation Benefits Scheme, power projects receiving funds from certain Central organizations, etc. In such cases, the Centre may have a greater degree of influence. This, however, begs the question of the extent to which the Centre's view of project planning is in fact different from that of the States.)

The River Boards Act 1956, passed by Parliament under Entry 56 of the Union List, provides only for the establishment of advisory boards, but no boards even of an advisory kind have been set up under the Act, and the Act has remained virtually inoperative. This point has been gone into in the discussion on 'basin planning'.

Other important Central enactments having a bearing on dam projects include the Environment Protection Act 1986 and the Forest Conservation Act 1980. Central clearances under these Acts are an essential part of the processes of approval of dam projects for inclusion in the national Plan. There is also the Wild Life (Protection) Act 1972 and the Water (Prevention and Control of Pollution) Act 1974. There are other water-related laws, e.g., legislation relating to groundwater, but they have only an indirect bearing on the subject being discussed in the present chapter.

There are numerous Acts at the State level (Irrigation Acts, Irrigation and Drainage Acts, and other related Acts)—and rules and regulations made under these—concerned with irrigation, canals and their maintenance, maintenance of tanks and so on. There are also Command Area Development Acts in several States. All these Acts have a bearing on the actual *operation* of dam projects. Currently there is some advocacy of separate legislation to provide a legal underpinning to the programme of Participatory Irrigation Management that seeks to transfer the management of irrigation systems below the outlet at a certain level to farmers' associations; the Andhra Pradesh Government has passed such an Act (the Andhra Pradesh Farmers Management of Irrigation Act 1997).

Dams are also built for the generation of hydroelectric power. Here the constitutional position is different. Electricity is in the Concurrent List, and both the Centre and the States can legislate (and consequently exercise executive power) on the subject. While irrigation (and 'multi-purpose') projects are undertaken by State Governments, there can be power projects in the Central sector as well. There are two Central corporations (the National Thermal Power Corporation (NTPC) and the National Hydro Power Corporation (NHPC) established for under-taking such projects. The Central Electricity Authority (CEA) is a statutory body established under The Electricity (Supply) Act 1948. Until recently, all power projects required a stautory techno-economic clearance by the CEA, but this has been substantially relaxed in the course of the economic reforms that have been going on since 1991. For facilitating private sector investments (both Indian and foreign) in the power sector, projects below certain capacity and investment levels have been exempted from the requirement of a clearance by the CEA.

Large-dam projects often involve the displacement, resettlement and rehabilitation of people. Taking displacement first, an important instru-ment of displacement is the Land Acquisition Act, dating back to the 19th century, under which private land is acquired by the state for a public purpose. This Act is invoked in the context of implementation of most large projects (including industrial projects), but it has a special importance for large-dam projects because of large land requirements. The actual operation of the Act has been beset with problems in many cases. Project planners and managers tend to complain about serious delays and protracted litigation, leading to slippages in project time-schedules and cost escalations; the people whose lands are being acquired tend to complain about inequities and injustices, disparities between cases, delays in compensation payments and corruption (apart from the upheaval involved in all displacement). The grievances of the people are clearly more weighty than the administrative inconveniences that project managers complain of. Though notifications are issued regarding the intention to acquire land, this does not really constitute consultation. The affected people can question the quantum of compensation, but it is very difficult for them to challenge the 'public purpose' claimed by the state, or to argue that alternative ways of achieving that public purpose should be considered. Until recently, there was no statutory requirement of a public hearing in relation to such projects. It has now been introduced, but has not yet become a well-established procedure. It is generally agreed that major changes are necessary in the Land Acquisition Act and related procedures.

In regard to both displacement and resettlement/rehabilitation, issues arise concerning equity, social justice and human rights, as between people and the state, as between people in the catchment or submergence areas and those in the command area, and as between different groups in the command area. Governments have indeed tried to provide project-affected persons (PAPs) with rights in the command area. Mention may be made of the Madhya Pradesh Project Affected Persons Resettlement Act (*Pariyojanaon ke Karan Visthapit Vyakti Punahsthapan Adhiniyam*) 1985; the Maharashtra Project Affected Persons Rehabilitation Act 1986; and the Karnataka Resettlement of Project Displaced Persons Act 1987. While these Acts are on the statute book and contain some enlightened provisions, it cannot be said that they have been fully put into practice. Similarly, well-intentioned provisions such as the collection of a 'betterment levy' from farmers whose lands get the benefit of irrigation at state expense, or a lower land ceiling for irrigated land as compared with unirrigated land, have remained largely unimplemented.

Apart from special legislation of the kinds mentioned above, the general laws of the country apply. Persons or groups who feel that they have been subjected to hardship or injustice can have recourse to the Fundamental Rights provisions of the Constitution and the writ juris-diction of the Courts, or to the National Human Rights Commission set up under the Protection of Human Rights Act 1993. Members of the scheduled castes and scheduled tribes could perhaps approach the National Commission for Scheduled Castes and Scheduled Tribes for support, though that institution may have no statutory role to play in this context. The Provisions of the Panchayats (Extension to the Sched-uled Areas) Act, 1996 (often referred to as the Tribal Self-Rule Act) could be invoked. UN conventions such as the ILO Convention 107 (ratified by India) have also a role to play. (Part IV of the Constitution, which lays down the Directive Principles of State Policy, is not justiciable, but those principles have to guide governmental actions; some of them have a bearing on dam projects.)

When projects are planned on rivers that cross India's borders and come from, or go into, other countries, the provisions of Article 51 of the Constitution (part of the 'Directive Principles'), as well as 'interna-tional law' in the sense of internationally accepted principles (earlier, the Helsinki Rules which had only the force of consent, and now the Convention on the use of international watercourses adopted by the UN General Assembly and awaiting ratification), have to be borne in mind.

India is not a signatory to the Convention and is not bound by it, but
if it does come into effect upon ratification by the required number of
countries it will acquire a certain moral force which will be difficult to
ignore.

We cannot leave the subject of laws without a reference to the no-
torious Official Secrets Act; this Act creates a veil of secrecy around
governmental actions, keeps the people at a distance, makes things as
difficult as possible even for individuals and NGOs with a proven record
of service to the people, hampers academic studies and in general renders
all talk of 'participatory' or 'people-centred' planning meaningless. In
cases of inter-State or international disputes, river flows are classified as
secret, making constructive work on conflict-resolution very difficult.
When certain projects face opposition and controversy on environmental or human (displacement) grounds, the Act is sometimes invoked to
deny information (or even physical access to places) to the people. This
is a widely recognized evil. There has been a movement for reform and
for a 'Freedom of Information Act', but this has not made much headway, though some States are formally ahead of the Centre in this regard.

We must take note here of the creation of case law by the Judiciary.
A new dimension has been added to the legal and policy framework by
the persistent efforts of the Judiciary to expand the human rights jurisdiction of the Courts and extend the scope of judicial review of
executive action, and its readiness to strike down laws that it considers
inconsistent with the spirit of the Constitution. This phenomenon,
known as 'judicial activism', is not without its critics, but it has been
widely welcomed in the country. As a part of this 'activism', the Judiciary
has been encouraging what is known as 'public interest litigation' (PIL).
Public-spirited individuals and NGOs have played an important role in
these developments. There are critics of PIL too, but it is generally
recognized that it has been a valuable innovation. Unfortunately, the
Supreme Court's judgement in the Narmada case marks a retrogression
(see Chapter 14).

Policy Framework

Once again, there is no special or separate policy statement on dam
projects, but there is what might be called an implicit policy in favour
of such projects. As already mentioned, the constitutional provisions
reflect an unconscious assumption that 'water' means canals and

storages. It is generally taken for granted that given the projected magnitudes of population growth and the concomitant needs of food, water and energy, and given the spatial and temporal variations in rainfall, large projects for the storage of river-flows and for the transfer of waters from surplus to deficit areas, or from one season to another, are necessary and desirable. That view, which finds wide acceptance, had its influence on the NWP 1987, and now stands enshrined in the NWP 2002.

Following the adoption of the NWP, some State Governments have formulated their own Water Policies (e.g., the Orissa Water Policy 1994, the Tamil Nadu Water Policy 1994). These take the NWP as the point of departure, reiterate some of its observations in different language and take some points slightly further. Unfortunately, they too remain largely in the realm of generalities. (Incidentally, both these documents assume that the Government will be the principal actor; this is explicit in the Orissa document, and implicit in the Tamil Nadu one.) While they do mention certain State-specific concerns, and contain some unexceptionable statements, they are essentially governmental exercises within the conventional framework, and mark no new approaches. In any case, it is too early to say what impact these documents will have.

Despite their limitations, the principles laid down in the NWP 1987 (and now 2002) on the subject of projects, if fully complied with, would undoubtedly lead to better project formulation and implementation. These are to an extent taken into account in the Guidelines issued by the Central Water Commission (CWC) for project preparation. Apart from this, successive National Plan documents contain observations on priorities for investment and sectoral policies. The September 1999 Report of the National Commission for Integrated Water Resources Development Plan (NCIWRDP) also contains observations on the planning, financing, implementation and prioritization of major water-related projects.

An important part of the policy framework relating to major/medium irrigation projects is the approach to the utilization of the irrigation potential that they create. An answer adopted in the 1970s was the Command Area Development Authority (CADA) system under which the state assumed the responsibility not only for creating a reservoir and a canal system, but also for taking the canal water closer to farms and for 'on-farm' development work. Along with this went extension and other activities relating to water management. Over a period of years, experience has shown that the CADA programme has been a limited success. There is now a move in a different direction, i.e., the state stepping back and handing over the management of systems below a certain level to farmers' associations, under what is known as 'participatory irrigation management' (PIM), to which reference has already been made.

Another important component of the policy framework (as distinguished from the laws referred to earlier) is the set of principles and practices that have been evolving on the subject of resettlement and rehabilitation of people displaced or otherwise affected by large projects. An effort to codify these in the form of a national rehabilitation policy has been in progress for some years, and various drafts have been under the consideration of the Government of India. The process does not seem to have reached finality so far. Meanwhile, the principles and practices adopted in particular projects, partly by the governments themselves and partly under the influence of NGOs and aid-giving agencies, constitute a body of precedents on the subject. The basic principle which has come to be generally accepted (but which does not necessarily mean that it has been fully or properly implemented in practice) is that PAPs should be so resettled and rehabilitated that their living conditions and 'quality of life' are at least as good as what they enjoyed before, and if possible, better; and further, that a certain minimum level is ensured for all, including those without title to land. Unfortunately, as mentioned earlier, there are serious deficiencies in the actual application of these principles in practice.

There are also other matters (such as agriculture, fisheries, environmental concerns, the welfare of scheduled castes and scheduled tribes, the welfare of women and children and so on) which fall within the ambit of different Ministries and agencies. On some of these matters there are formal Policy Statements or Laws (e.g., the National Policy on the Environment and the National Forest Policy, apart from the Environment Protection, Forest Conservation and Tribal Self-Rule Acts already mentioned). On others there are sectoral policies, decisions, etc., adopted by the respective Ministries.

For instance, there is the 'Policy on Hydro Power Development' brought out by the Ministry of Power (MoP) in August 1998. After a very brief general statement of the familiar case for hydropower projects, the document quickly proceeds to the means of achieving the substantial capacity additions felt to be needed. There is no real policy analysis in the sense of exploration of options and alternatives. Possibilities such as a combination of demand management, energy-saving, getting more out of capacities already created, extensive decentralized generation and so on, with minimal recourse to large projects, are not discussed.

On uses other than irrigation and hydroelectric power (e.g., navigation, industrial uses, municipal uses, etc.), the appropriate Ministries/Departments (Surface Transport, Industry, Rural Development, Urban Development, etc.), have their policies and concerns. These need to be kept in mind by the planners of water resource projects.

All these (i.e., laws, policy statements, implicit policies, administrative decisions, case law, etc.), together constitute the policy framework with reference to which projects have to be prepared, approved and implemented.

Planning System

The Indian planning system and the complexities of plan finance cannot be set forth in detail here. We need merely note:

- that there is a system of centralized economic planning (not of the 'indicative' kind, but concerned with targets and projects, approving financial outlays and so on, and virtually performing some of the functions of the Finance Ministry, though the on-going economic reforms have begun to reduce the relevance of such planning);
- that this is managed by the Planning Commission, a non-statutory body which derives its influence from the role it plays in the determination of Plan outlay levels, the allocation of outlays for projects and programmes, and Central Plan assistance to the States;
- that the Planning Commission is concerned with planning at both Central and State levels;
- that the planning is for five-year blocks broken down into annual plans to correspond to the annual budgeting;
- that the planning is essentially for the public sector; and
- that all developmental projects must form part of the Plan of the State concerned.

Once a dam project is included in the State Plan in accordance with a procedure that will be described later in this chapter, it is funded out of the provision made in the State budget on the basis of the outlay approved for it in the national Plan. It then qualifies for inclusion in the calculation of Central Plan assistance to the State Plan in accordance with the Gadgil Formula referred to earlier.

(Note: In addition to Plan assistance transfers from the Centre to the States, there is also a redistribution of revenues between the Centre and the States under the recommendations of quinquennial Finance Commissions as mandated by the Constitution, and some 'non-plan' forms of Central assistance. These matters have a bearing on the subject of this chapter in so far as provisions for the operation and maintenance of projects are concerned. The division of governmental budget provisions and expenditures into 'plan' and 'non-plan' categories is problematic and has been much criticised. Outlays on projects are classified as 'plan' and operation and maintenance as 'non-plan'; and there is an oversimplified view that in a situation of a scarcity of resources, plan expenditures merit priority over non-plan expenditures. This, together with the inherent tendency on the part of the bureaucracy, technocracy and politicians in India to be more interested in the implementation of new projects rather than in the efficient running of completed projects, leads to the under-provisioning and neglect of maintenance.)

In recent years there has been some talk of 'private sector participation' in large-dam projects. This has made only modest beginnings even in the case of thermal power projects so far, and much more so in the case of hydropower and/or irrigation projects. As and when this becomes a significant reality, a crucial question will be how the human and social aspects, which have presented great difficulties even in public sector projects, will be taken care of in private ones. (It must be noted that the NWP 2002 incorporates a new paragraph on private sector participation.)

Project Planning, Approval, Implementation: Institutions, Procedures

Identification of a Project

Theoretically speaking, projects can get identified in several possible ways:

  (i) a consideration of the needs of a particular area or region (drinking water, irrigation, water for municipal/industrial uses, hydroelectric power, flood management, etc.,) could lead to the postulation of a project; or

(ii) from an engineering point of view, a particular location may be identified as a suitable place for the construction of a dam and the creation of a reservoir (or a barrage or a weir, as the case may be); or

(iii) an overall master plan or outline plan for an area (a basin or sub-basin) may envisage a number of projects at various locations.

Instances of all three kinds of projects can be mentioned. For instance, the Telugu Ganga Project was initially conceived as an answer to the water needs of Madras city (now Chennai), and therefore it can be classified as an example of case (i). Again, many of the projects that have been undertaken over the years have come up in sites that had been identified long ago, or those that were determined in an early exercise of the erstwhile Central Water and Power Commission under which the resources of various basins were assessed and locations for storages indicated; these may therefore seem to fall into case (iii). However, the distinction is in a sense illusory. Once it is assumed that 'water resource development' is a matter of an engineering intervention—and this assumption is common to all three cases—topography and geology become important, and suitability from an engineering point of view becomes the determining factor in the identification of projects.

It may be felt that some projects emerge out of the processes of electoral politics and acquire a hold over the popular mind, and that this is one more category that must be added to the three above-mentioned cases. However, the general public and politicians do not invent projects; they generally pick up an idea that is in the air. That idea must have initially been thought of by a person with some technical background, however tenuous. If so, this does not constitute a new category.

Case (iii) leads to the question of the extent to which 'basin planning' has been actually practised in India. This has been discussed elsewhere in this book, and that ground need not be covered here again.

Preparation, Examination and Approval of Projects

The basic agency for the preparation of a large-dam project is the Irrigation or Water Resources Department of the State Government concerned, which is largely a department of engineers. The various non-engineering aspects—agricultural, environmental, energy, financial, economic, social, etc.,—are partly dealt with departmentally and partly through consultations with and comments by the agencies and departments concerned. In preparing a project, the Guidelines of the CWC and of the Ministry of Environment and Forests (MoEF) need to be kept in mind (as the project will eventually go to these agencies for examination, as explained earlier). A project so prepared goes through the usual approval procedures within the State Government (e.g., consultation with the Finance, Planning and other Departments concerned, and with the State Planning Board, if any).

After approval at the State Government level, the project goes to the Central Government for approval. As already mentioned, there is no clear constitutional or legal prescription of approval of such State projects by the Central Government, but the requirement has become established in two ways: first, as a condition for the acceptance of a (major or medium) dam project for inclusion in the National Plan; and secondly, as a requirement of clearances by the MoEF under the Environment Protection Act and under the Forest Conservation Act. A third dimension is added by the fact that in respect of inter-State rivers, i.e., rivers running through more than one State, there is an inter-State angle which has to be attended to by the Central Government.

The institutional machinery for the techno-economic clearance of major/medium irrigation or multipurpose projects at the Centre is the Advisory Committee on Irrigation and Multipurpose Projects (popularly though misleadingly known as the Technical Advisory Committee or TAC). This is a Committee set up by the Planning Commission. It is chaired by the Secretary, Ministry of Water Resources (but is not a Committee of that Ministry). The TAC is serviced by the CWC which does the detailed techno-economic examination of projects. (Incidentally, the CWC is not a statutory body like the CEA.)

The project sent by the State Government to the CWC as the Secretariat of the Technical Advisory Committee (TAC) is examined in the different Directorates of the CWC and other agencies concerned are consulted. Queries are raised, suggestions are made, changes are proposed and so on, and when this process is complete, the CWC submits a note to the TAC. Representatives of other concerns (e.g., agriculture, environment, energy, finance, the welfare of scheduled castes and scheduled tribes, women's issues, etc.,) are expected to participate in its meetings. In the case of inter-State rivers, the relevant inter-State issues are an important part of the examination. At the TAC the discussion proceeds on the basis of the CWC's note. If the project is found acceptable, the techno-economic clearance of the project by the TAC (i.e., a judgement as to whether the project is technically sound and economically viable) is communicated to the Planning Commission.

Separately, two clearances by the MoEF (as mentioned above) are also needed; one under the Forest Conservation Act and the other under the Environment Protection Act. (The latter was earlier a requirement laid down by policy, but is now statutory.) For the purposes of these clearances, the MoEF is assisted by two separate committees, namely, the Environmental Appraisal Committee and the Forest Advisory Committee. These Committees have their own procedures, and after these are completed, they make their recommendations to the MoEF. If the recommendations are favourable, the MoEF issues two letters of clearance; one relating to the environmental aspects, and the other relating to clearance under the Forest Conservation Act, with conditions attached, if necessary.

After the techno-economic clearance (including inter-State aspects) by the TAC and the clearance by the MoEF from the environmental and forest angles, the Planning Commission examines the project from the point of view of investment priorities, sectoral planning policies and provision of funds in the national Plan, and issues a letter of acceptance (often referred to as 'investment approval'), again with conditions attached, if necessary.

Thereafter, the implementation of the project is the responsibility of the State Government concerned. Save in exceptional cases where a public sector corporate or autonomous agency is created for the implementation of a project (e.g., Tehri, Sardar Sarovar), or the project is entrusted to a public or private sector company (e.g., projects under implementation by the NHPC, the Maheshwar Project in Madhya Pradesh entrusted to a private corporate body), the implementation of a dam project is generally a departmental responsibility, carried out through consultancy/construction contracts. (As mentioned earlier, the idea of private sector participation in such projects has made only limited headway so far.)

[Note: The procedure outlined above applies to all major/medium irrigation and multipurpose projects that come within the purview of the Ministry of Water Resources (MoWR). Hydroelectric projects proper (i.e., without a significant irrigation component) are the concern of the

MoP. Further, irrigation and multipurpose projects are essentially State projects that come to the Centre only for inclusion in the national Plan (apart from certain statutory clearances), whereas in the field of power there are also Central projects for which the administrative ministry is the MoP. For instance, the Sardar Sarovar Project (Gujarat) and the Narmada Sagar Project (Madhya Pradesh) are State-level multipurpose projects and are the concern of the MoWR, whereas the Tehri Hydro-Electric Project, essentially a power project though it has a minor irrigation component, is a joint project of the Central and Uttar Pradesh Governments which is being implemented by a joint corporate body that is under the administrative control of the MoP. Power projects (including hydroelectric projects) falling within the purview of the MoP go not to the TAC but to the CEA for a techno-economic clearance. In such cases, the 'hydro' aspects of the projects are examined by a Member (Hydro) in the CEA and the civil engineering staff supporting that Member. That examination is similar to the CWC's examination of projects for the TAC, and is based on common traditions inherited from the days when there was a joint Central Water and Power Commission. After the techno-economic clearance by CEA, power projects too (whether thermal or hydro) need clearances under the Environment Protection Act and the Forest Conservation Act by the MoEF, and an approval by the Planning Commission for inclusion in the national plan. However, single-purpose hydroelectric projects are few in number; by and large, most dam projects go through the TAC route. In any case, dam projects, whether for irrigation or hydroelectric power, involve the same problems of displacement, environmental impacts, etc., and suffer from the same limitations and deficiencies.

Investment Criteria

The TAC's basic criterion for the approval of projects has been the benefit-cost ratio (BCR). In the colonial period, irrigation projects were often sanctioned for the purpose of earning revenue, and the criterion for sanction was the rate of financial return to the Government on the investment. With the advent of economic planning in independent India, it was felt that such projects should be judged not on the basis of the revenues accruing to the exchequer but on that of the benefits accruing to the economy. Accordingly, the BCR was adopted as the investment criterion instead of a financial return. The BCR does not imply a complex and sophisticated socio-economic cost-benefit analysis. 'Cost' in this context meants merely the direct cost to be incurred on the project, and 'benefit' meants the incremental agricultural production expected as a result of the irrigation provided by the project, and/ or the value of the power to be generated. The BCR is calculated for an average year after project completion: discounted cash-flow calculations are not undertaken. A BCR of 1.5:1 was considered desirable, but lower standards (1:1 or even less) were considered acceptable for projects in drought-prone areas.

In the case of industrial or commercial projects, a fairly sophisticated techno-economic cost-benefit analysis using concepts such as opportunity costs, shadow prices, etc., and calculating both financial and economic internal rates of return (IRR), has been in vogue since the early 1970s. This analysis is carried out by the Projects Appraisal Division (PAD) of the Planning Commission and is submitted to the Public Investments Board (PIB) of the Ministry of Finance to which all such projects need to go for clearance. The PIB procedure (which applies also to Central power projects dealt with by the MoP) does not apply to major/medium irrigation and multipurpose projects of State Governments which come to the Centre for a clearance for inclusion in the national Plan. These projects, as already mentioned, go to the TAC where the criterion for their approval is a rather simple BCR. The BCR as actually operated is an unsatisfactory criterion, and one which is liable to distortion. It is often stated with some plausibility (though it is difficult to substantiate this) that costs are deliberately understated and benefits overstated in order to arrive at a BCR of 1.5:1. Dissatisfaction with the manner in which irrigation and multipurpose projects were being dealt with led to the establishment of a Committee by the Planning Commission to go into this matter. The Committee (the Nitin Desai Committee) submitted a report in 1983 making recommendations for a changeover to a better appraisal system involving a proper socio-economic cost-benefit analysis leading to the determination of an economic IRR. This remains unimplemented. The BCR has indeed been replaced by an IRR, but the elements that go into this are the same as before, and there is no real socio-economic cost-benefit analysis. The new IRR is only the old BCR stated in a different way. The NCIWRDP (1999) has gone into this matter and reiterated the need for a move towards the kind of appraisal that the Nitin Desai Committee had recommended.

Partly as a consequence of the abandonment of the financial return criterion, and partly because the pricing of irrigation water in many States is so low and the recovery so poor as to make it virtually free, most large-dam projects are loss-making propositions in so far as the state exchequer is concerned. This has been well brought out in the Report of the Committee on the Pricing of Irrigation Water (the Vaidyanathan Committee, 1992, Planning Commission). This aggravates the resource shortage of the States and makes the proper maintenance of systems and the provision of a satisfactory service to the people even more difficult. This is compounded by the Plan/non-Plan distinction and the innate preference for new projects over the maintenance of existing ones, to which a reference was earlier made.

It needs to be mentioned that projects posed to the World Bank or other external aid-giving agencies go through a process of detailed examination by that agency. There have been criticisms that the donor agencies' concerns and conditions tend to exert an undue influence on, and in some cases distort, internal policies and priorities; that aided projects tend to become 'gold-plated' in respect of certain features and heavily capital-intensive; that they receive a priority in budgetary allocations to the detriment of other projects and activities; that the very fact of project-lending builds a bias in favour of 'projects' to the disadvantage of alternative approaches, and so on. At the same time, it is perhaps true to say that the questions raised and requirements stipulated during the appraisal and negotiation processes, and the conditions attached to the loan, have in some cases contributed to improvements in project planning as also in environmental and rehabilitation measures.

Some Weaknesses

Some infirmities of the aforementioned systems and procedures may be noted here.

(i) The primary and controlling discipline in project preparation at the State level is engineering. Projects are prepared by the State's Irrigation Department (or to use a more recent nomenclature, the Water Resources Department), which is primarily a department of engineers. Other disciplines, concerns and points of view are to some extent brought in through consultations and comments, but there is no *interdisciplinary* planning in the proper sense of the term. At the Centre, too, the primary project-examining body, i.e., the CWC, is mainly a body of engineers. Here again, other concerns and disciplines are brought in through consultations and comments, but there is no interdisciplinary approach. The CWC has some non-engineering personnel on its staff (though not at very senior levels), but the Commission itself has only engineer-members, and it is mainly staffed by officers of the Central Water Engineering Service. Thus, it is a Water Engineering Commission rather than a Water Commission, though it is often referred to as the 'apex' expert body on water resources in the country. A High Level Committee to review and restructure it set up in 1995 was in the process of discussing the possibility and desirability of making the CWC a multidisciplinary body and a true water commission, but for certain reasons the Committee became defunct before it could complete its work. The NCIWRDP (1999) has now recommended that the CWC should be transformed on those lines.

(ii) It follows from the above that there is and can be no integrated 'holistic' planning under these circumstances, despite instructions and exhortations to that effect. For instance, environ-mental considerations are supposed to be fully 'internalized', but this has not begun to happen to a significant extent. This aspect is still largely looked upon as an externally imposed discipline that has to be complied with. An 'environmental impact assessment' (EIA) has no doubt been a prescribed requirement for all projects for some years and is being complied with, but EIAs are not entirely satisfactory in many cases, as has already been pointed out. That is why the examination by the MoEF often results in basic questions being raised at a late stage, leading to complaints of projects being delayed. What this points to is the failure to internalize and integrate these concerns into project planning from the earliest stages.

(iii) One major feature of project planning has been the dominance of irrigation. Even 'multipurpose projects' often have only two components, namely irrigation and hydroelectric power. The integration of other purposes, such as the provision of drinking water, flood-moderation, navigation, the maintenance of downstream flows and so on, has not been a standard feature of project planning. There could be conflicts between two different uses (e.g., between irrigation/power generation and flood-moderation, between irrigation and maintaining minimum flows), but these are not always explicitly recognized and built into project planning.

(iv) The next point, which is again related to the points made above, is that only one unique project is placed before the TAC for approval. The TAC note of the CWC gives a technical account of the particular project that is submitted for approval, and proceeds to examine it. It follows that project decisions do not represent carefully considered choices out of a number of possible answers to a given need or problem. *Within* the ambit of a project, there may be multiple possibilities at various stages. Some of these are purely technical or engineering choices, and these are likely to be covered in the processes of project preparation and in the TAC Note. However, complex techno-socio-ecological-economic choices are not usually considered. For instance, there is no conscious principle or policy of *'minimum environmental impact'* or *'least displacement'*, and choices based on such considerations are unlikely to be presented.

'Alternatives' to a given dam project could include:

- extensive efforts at local water-harvesting and/or watershed development;
- getting more out of capacities (say, for power generation) already built, in respect of which 'social costs' have already been incurred;
- adding to the capacity of an existing project (say, a reservoir) through supplementary investments instead of undertaking a new project;
- minimizing the need for supply-side solutions through demand management, increased efficiency of resource-use, resource-conservation, etc., and so on.

'Alternatives' could also mean a *combination* or *integration* of a large project with smaller projects and/or with watershed development programmes in a holistic plan for an entire area. Alternatives in these senses are not usually considered.

(v) One of the factors that militate against holistic, integrated planning is the fragmentation and compartmentalization of responsibilities at the administrative level. The distribution of subjects among different Ministries/Departments, whether at the Centre or in the States (for instance those dealing with water resources, hydropower, agriculture, fisheries, navigation, rural development, urban affairs, environmental maters and so on), is doubtless inescapable, and no one will seriously argue for the clubbing of all matters in which water has a role to play into one Ministry/ Department. However, even within the area of water resources proper, there is a compartmentalization of different components or aspects such as major/medium projects; minor irrigation; command-area development; groundwater; watershed development; rainwater harvesting; water management, and so on. Different Divisions/Departments/Agencies tend to deal with these matters with little coordination, much less integration.

(vi) There is reason to believe that the appraisal and decision-making at the TAC level is not rigorous enough. The basis for this remark is partly that too many projects have sometimes been cleared at one meeting of the TAC, and partly that the post-clearance history of scope changes and modifications is a reflection not only on the quality of project planning but also on that of appraisal and approval. Further, with the kind of investment criterion employed, as already discussed, the appraisal cannot possibly be rigorous.

(vii) Civil society (in the sense of the people concerned, i.e., beneficiaries and those who are likely to be adversely affected, and the community in general) plays little or no role in the planning and implementation of such projects. The activity is essentially governmental. Project planning and implementation are largely internal activities of the state. As the colonial state had consciously distanced itself from the people, and as that distance did not significantly narrow in the post-colonial era, a tradition of consultation of, and participation by, the people did not develop. It is only in recent years that a consciousness of the importance of what has come to be called 'stakeholder participation' has begun to emerge. PAPs, with the assistance of some NGOs, have become more conscious of their rights (both their fundamental rights as citizens and their traditional rights to the use of river waters, forest produce and other natural resources). The Government for its part is trying to formulate a national rehabilitation policy. Public hearings have now become a statutory requirement (but we have still to see how this evolves in practice).

(viii) In the absence of institutional arrangements for consultation and grievance-redressal, the processes of displacement, resettlement and rehabilitation often generate serious dissatisfactions. When these lead to popular resistance or agitation under the leadership of NGOs, the state machinery tends to respond with incomprehension and sometimes force. This has happened in the case of several projects. Bureaucratic traditions do not facilitate a good working relationship between the Government and NGOs. The state in turn tends to charge NGOs with adopting a confrontationist attitude and hampering the activities of the state. However, it is often the state which, through its failures of consultation, delays, reluctance to part with information and woodenness and unimaginativeness in implementation, forces the people and the NGOs on to the confrontationist path.

(ix) In the context of the planning and funding processes, four main (interrelated) dysfunctional features relating to major irrigation/multipurpose projects need to be noted: (*a*) the thin and sub-optimal spreading of resources on a large number of projects; (*b*) the time and cost overruns on many projects; (*c*) the persistent problem of projects remaining forever incomplete, spilling over several Plan periods, and pre-empting Plan resources for continuance/completion, leaving hardly any funds for new projects; and (*d*) the failure in many cases to achieve the projected benefits in full measure. Successive Plan documents have stressed the need for better project planning and implementation and for complet-. ing ongoing projects before starting on new ones, but to little purpose. From the Sixth Plan onwards the theme has been 'consolidation' and 'no new starts', but this has not been effective. These matters have been discussed at some length in the Report of the NCIWRDP (1999).

(x) The monitoring system is weak, and there is no effective mechanism to ensure that wherever sanctioned costs are likely to be, or have been, exceeded, the revised cost estimate (RCE) is promptly brought to the TAC/Planning Commission for a fresh appraisal. When at last the RCE comes before these bodies the reappraisal serves little useful purpose, as the option of reviewing, and if necessary reversing, the investment decision already taken no longer exists; at best only minor changes or adjustments may be possible at that stage. There is also no established system of a post-completion evaluation. Very few projects, other than those that receive World Bank assistance, are subjected to such an *ex post facto* reappraisal.

(xi) Where the approval of a project is conditional, there is no effective mechanism for ensuring compliance with those conditions and taking appropriate measures in the event of non-compliance. This is clear enough from the Reports of the Five Member Group on the Sardar Sarovar Project and the Expert Committee on the Environmental and Rehabilitation Aspects of the Tehri Hydro-Electric Project. (The *pari passu* condition imposed in these cases has been misinterpreted, and has not really worked, as was pointed out in Chapter 5, section I).

References to many of the failings mentioned, couched perhaps in different terms, will be found scattered in various documents, such as the successive Five Year Plans and the Reports of numerous Committees and Commissions.

14

The Narmada Judgement

A Disquieting Judgement

This chapter is about the nature and implications of the Supreme Court's judgement of October 2000 on the Narmada (Sardar Sarovar) case, and not about the merits of the project or about the question of large dams in general. 'Judgement' here refers to the majority judgement by Justices Kirpal and Anand. While the minority judgement by Justice Bharucha is not without importance, it is the majority judgement that prevails and constitutes *the* judgement in this case; and being the judgement of the highest court in the land, it represents *finality* from a legal point of view. A review petition submitted by the Narmada Bachao Andolan (NBA) failed and the judgement was reaffirmed. Nevertheless, a critique of the judgement has seems not merely warranted but necessary for reasons that will become clear enough as this chapter proceeds.

During the last decade or two, the Supreme Court has been blazing a trail. While there has been some criticism of what has come to be known as 'judicial activism', it has on the whole won national approval. Most people (the author included) have been grateful to the Judiciary for trying to rescue the country from the egregious failures of the Executive and the Legislature. Unfortunately, all that good work was nullified at one stroke by this single judgement, which blazed a trail in the wrong direction. The complaint of the author is *not* that the judgement allowed the project to proceed further. It was never the author's expectation that the Court would stop the project. However, he *had* hoped that approval to further construction would be severely conditional and that justice would be done to the project-affected persons (PAPs). Those hopes were belied. The judgement can only be described, with deep regret, as a most unfortunate and disquieting one. Such a statement cannot be made lightly; the following sections will provide the necessary justification.

Failure to Deal with Issue

First, the judgement, delivered after six long years of proceedings, failed to deal with the very issue that was brought before it, namely, a situation of lapse and failure in relation to certain aspects. The judgement allowed the dam to go up from the height of 85 metres where it had been stopped for some time to 90 metres (in fact it wanted this to be done expeditiously), and stipulated that further construction would be conditional on clearances by the Environmental and Rehabilitation Sub-Groups of the Narmada Control Authority (NCA) from their respective points of view and with reference to the conditions of clearance with which they were concerned. Those Sub-Groups were in any case charged with that responsibility, and the judgement has said nothing new; this chapter will return to that question. However, the point that needs to be noted here is that if a check with reference to the environmental and rehabilitation aspects was warranted after 90 metres, it was equally warranted *before* that height was reached. When the judgement was delivered, the rehabilitation of PAPs had not been completed fully even in relation to a height of 85 metres. This must have been clear enough from the material before the Court. It had also been clearly stated that land for resettlement was not available in Maharashtra and Madhya Pradesh. The judgement itself faults the Madhya Pradesh Government for its failures in this regard. There were deficiencies in relation to the environmental conditions too. It was beyond doubt that the *pari passu* clause had not been complied with. Thus, there was an *existing* situation of failure of compliance with the conditions prescribed by the Ministry of Environment and Forests and the Planning Commission while according approval to the project in 1987. That failure also constituted a violation of the Narmada Tribunal's directions as well as those given by the Supreme Court itself in the past. The minority judgement of Justice Bharucha did not specifically refer to this, but in a sense it went further: it said that the very clearance given to the project in 1987 was wrong because it was not based on a proper examination; on that ground it called for a halt to the project until it was put through a fresh scrutiny and clearance. There was thus room for some concern (to put it mildly) on the environmental and rehabilitation fronts. Justices Kirpal and Anand might not have agreed with Justice Bharucha that there was need for a fresh examination and clearance, but should they not at least have made further progress from 85 metres to

90 metres conditional on the existing deficiencies being remedied and compliance completed? Overlooking present non-compliance and asking for compliance to be checked at some future time amounted to a condonation of violations—a kind of 'amnesty' scheme for the project authorities and the Governments concerned. (It might be argued that the Rehabilitation Sub-Group of the NCA had found that the conditions had been fulfilled for a height of 90 metres. They had said nothing of the kind; they had merely noted that 'arrangements were in place', and not that conditions had been fully complied with. Nothing that they said could have been construed as warranting further construction up to 90 metres.)

Essay on Dams

Secondly, the judgement muddied the waters by making sweeping pronouncements about the desirability of dams. Curiously enough, the learned judges began by arguing the case for judicial restraint and chastising the petitioners for bringing before the Court matters that belonged to the executive sphere. They seemed to have forgotten that they themselves, or their predecessors on the Bench, had raised some issues of this kind (hydrology, height of the dam, etc.,) and asked for a second Report from the Five Member Group (FMG), and that the further Report of the Group had been submitted in April 1995 not to the Central Government, but to the Supreme Court as directed by it. Leaving that aside, and accepting the stress on judicial restraint as valid, one must ask why the learned judges then proceeded to write an essay on the virtues of dams. The petitioners were not asking for an injunction against dams in general; they were in fact instructed by the Court at an early stage of the case not to raise general issues regarding dams but to confine themselves to the particular project in question. They were not even asking for an immediate abandonment of the Sardar Sarovar Project (whatever their views on dams in general and this dam in particular might have been) but making submissions on what they considered to be the adverse environmental, social, human and economic consequences of the project, and asking for a stoppage of work on the project pending a comprehensive independent review. Conceivably, such a review could have led to either a negative or a positive conclusion. Speaking subject to correction, nothing in the petitioners' submissions called for an Ode to Dams by Their Lordships.

Apart from that inconsistency, the advocacy of dams in the judgement is undistinguished, to say the least. The judgement puts forward the familiar arguments for dams (variability of rainfall over time and space, need for storages and transfers, the 'clean' nature of hydroelectric power, etc.,); dismisses the advocacy of alternatives such as watershed development and local water-harvesting; avers that dams are necessary for development (with no hint of an awareness of the debate regarding the notion of 'development'); makes light of the adverse impacts of dams; goes to the extent of saying that dams are actually good for the environment; seems to accept the doctrine that some people must 'sacrifice' (be sacrificed?) for the good of others, and observes that no instance of a dam having done any harm has been brought to notice! One keeps rubbing one's eyes in disbelief that the learned judges could really have made themselves responsible for such rash and ill-considered statements. If this had been a presentation made in a seminar or conference, it would have received short shrift. The case for dams could have been much better argued; and the case against dams can be argued with even greater force. There is a major unresolved controversy and a vast literature on this subject. There are books on large dams by Goldsmith and Hildyard, Patrick McCully, B.D. Dhawan, Enakshi Thukral Ganguly, Jean Dreze et al., Satyajit Singh and many others. There was also the India Country Report (*Large Dams: Indian Experience*) submitted by a team of five to the World Commission on Dams (WCD). (When the judgement was written, the WCD's own Report was expected to be released soon.) Against that background, one wishes that the judgement had not rushed headlong into this dangerous terrain. In any case, the mere fact that these observations about dams have been made by judges in a judgement does not give them any greater legal force than the views of others—engineers, economists, sociologists, environmentalists or even ordinary people. They remain mere unsupported personal opinions that have no place in a judgement.

Disapproval of Narmada Bachao Andolan

Thirdly, apart from a naïve belief in the virtues of dams (and an undercurrent of disapproval directed at those who argue against them), another force driving this judgement is a strong disapproval of the NBA. Consider some of the remarks about the NBA made by the learned judges: 'an anti-dam organization'; 'publicity interest litigation', 'private inquisitiveness litigation', and so on. With respect, these dismissive remarks do not reflect a judicious frame of mind. The petitioners managed to persuade at least one judge that there was something in what they were saying; this was therefore hardly a case of frivolous or trivial public interest litigation (PIL). Besides, the submissions made and documentation presented could have been accepted or rejected by the Court, but nothing in them could have warranted the kind of remarks cited above. The animus that is evident in the judgement is very similar to the anger that marred the judgement in the famous Election Commission (T.N. Seshan) case some years ago.

Incidentally, the very use of the expression 'anti-dam' with an undertone of disapproval is revealing. Why should that term carry a pejorative connotation? Both 'pro-dam' and 'anti-dam' positions are surely mixtures of valid and invalid arguments?

Laches Argument

The animus against the petitioners is particularly evident in the section entitled 'Laches' in which strictures have been passed on the NBA for its delay in bringing the case to the Court. The point made is that the clearance to the project was given in 1987, whereas the NBA came to the Court as late as 1994, on which ground alone, according to the learned judges, the petition could have been rejected. With respect, one wishes that they had done so; much time would have been saved, and the NBA would have been left free to explore other channels or forums. The answer to the charge of delay is evident and was available in the material before the court. The NBA started by trying to improve the rehabilitation policies and packages and their implementation. Over a period of time it gradually came to the conclusion that the project was badly flawed and needed a major review. It was only at that stage that the NBA began to think of going to the Supreme Court, partly encouraged by the new receptivity of the Court. Meanwhile the NBA's campaign produced some results: the World Bank appointed an Independent Review, and some time later, the Government of India set up the FMG. Unfortunately, the Gujarat Government boycotted the FMG and questioned its constitutionality, and someone filed a case in the Gujarat High Court against the establishment of the FMG. It was due to a growing sense of despair at the failure of its efforts to find an adequate response from the executive machinery that the NBA thought that it should move the Supreme Court. That is an understandable development and one fails to see any occasion for a reprimand. It must be noted that Justice Bharucha finds no merit in the 'laches' argument.

It may not be out of place here for the author to narrate a piece of relevant history. When the NBA brought the FMG's Report of April 1994 to the notice of the Supreme Court, the judges wanted a supplementary report on certain aspects and wanted to know how much time the FMG would need. This was in March 1995. The FMG said that it would give a report by 31 May 1995. Their Lordships were furious. They asked the Solicitor General whether the members of the FMG did not realize the urgency of the matter, and said that they wanted a report by 16 April 1995 as they were anxious to pass an early judgement on the case. Some members of the FMG were unhappy at the tone of the judges' remarks in Court, but out of deference to the Apex Court the Group agreed to do whatever it could; and its Report was submitted to the Court on 16 April 1995 as required. The Court delivered its judgement in October 2000. Presumably it would be improper to ask why the Supreme Court took so long to pass a judgement in this case, but against that background the learned judges could at least have refrained from talking about the petitioner's putative delay.

Existing Institutions Presumed to be Working

Fourthly, in allowing construction to proceed and asking for checks to be made after the height of 90 metres has been reached, the judgement introduced no new safeguards to ensure compliance but advanced the doctrine that the existing institutions must be presumed to be working. That doctrine was not corroborated by actual experience. The judgement referred to the NCA, its Environment and Rehabilitation Sub-Groups, and the ministerial-level Review Committee. All this was in existence already. The *pari passu* clause implied a continuous check to see that construction did not proceed ahead of measures on the environmental and rehabilitation fronts, but it had broken down. The Environment and Rehabilitation Sub-Groups had not been very effective. In this context, the author (who was also a member of the FMG) would like to state the following. During the course of a session that the FMG had with a former secretary of the Environment Ministry and Chairman of the Environment Sub-Group, that functionary expressed his anguish at the difficulties that he had experienced in discharging his responsibilities, the inadequate response that he had received from his colleagues on the Sub-Group, and the peer pressure on him to be 'positive' and not stand in the way of construction, and said that he wanted the FMG to take note of this. There is a veiled reference to this in the FMG's Report. That very passage was cited in the judgement but without an appreciation of its significance.

The failure of the existing machinery was in fact what led to the PIL. The NBA must be presumed to have established some kind of a *prima facie* case, because the Court itself suspended construction for several years, and finally at least one judge found enough ground (in the minority judgement) for ordering a fresh scrutiny and clearance. What then was the point in recapitulating (in the majority judgement) the existing arrangements as if new orders were being passed? Even the prescription of a reference to the Prime Minister (as if he were a judicial authority), inappropriate as it was, was nothing new. The Review Committee is a ministerial-level Committee. A disagreement at that level is bound to lead to a reference to the Cabinet or to the Prime Minister; this has happened before. The judgement offered nothing new. It was in fact a denial of relief and of justice.

A Volte-Face

The Court has not in the past presumed that the existing machinery is working. One of the innovations of the Indian judiciary has been the assumption of the right to ask public authorities why they have not been discharging their statutory responsibilities. The Court has given directions to the Central Bureau of Investigation in certain cases, and asked for periodical reports; it has gone into garbage clearance by municipalities; questioned public health authorities on measures to prevent the outbreak of dengue; directed the shifting of industries; laid down schemes for admissions to educational institutions in the private sector; expressed displeasure with pollution-control measures; concerned itself with the state of the Yamuna river, with the saving of the Taj Mahal from the effects of pollution and with emission norms for automobiles, and ordered the establishment of the Central Groundwater Authority. Some of this was undoubtedly excessive activism, but the presumption in the Narmada judgement that existing arrangements were working was a complete *volte-face*.

The steady widening of the scope of judicial review during the last several years (sometimes carried too far) yields place in this judgement to the doctrine of 'Government knows best' and the abdication of judicial responsibility for protecting the rights of the people; the earlier encouragement of PIL (again, sometimes carried too far) changes to the deprecation of PIL in sarcastic language; and the enthusiasm for environmental causes (not always well thought out) is replaced by faith in government committees and the proposition that these are not matters for the Courts. This was indeed a full-scale retreat on a wide front. Assuming that some degree of correction had to be applied to the excesses of judicial activism, the learned judges need not have gone quite so far; and it was ironic that for the purpose of cutting PIL down to size they should have chosen the one case where, more than in any other, PIL was appropriate and called for.

Denial of Justice

This could not even be regarded as 'passivism' as opposed to the earlier 'activism'; it was in fact activism of the wrong kind—on behalf of the state. In future, if the existing machinery fails and if the state uses the police as an instrument of enforcement of its policies for what it considers 'development', what recourse does the citizen have? In effect the Narmada judgement threw the affected people back to the tender mercies of the governmental machinery. The people who had approached the Supreme Court for justice were given a dusty answer. (This invites comparison with the judgement that upheld the suspension of Fundamental Rights during the infamous Emergency.) This was a severe setback not just for the NBA and Medha Patkar but to all movements for the empowerment of the people vis-à-vis the state and the cause of environmental protection.

The 'Non-Reviewability' Argument

Finally, something needs to be said on the 'non-reviewability' of certain portions of the Tribunal's award, as the judgement seems to set much store by that argument of the Gujarat Government. A Tribunal is essentially a conflict-resolution mechanism. Its award (including the 'non-reviewability' of parts of it) is indeed binding on the parties to the dispute in the sense that no party can unilaterally resile from it. However, if *all* the parties to the dispute reach an agreement, surely they can not only make changes but even set aside the Award and sign a new accord. Further, in the context of the Inter-State Water Disputes Act (ISWD Act) 1956 an 'inter-State dispute' means essentially an *inter-governmental* dispute: if it has been resolved without consulting the people whose interests are affected, can they be asked to accept the consequences, say, a project involving displacement on a large scale, without demur? Is an Award under the ISWD Act also an adjudication between the state and the people (who were not parties before the Tribunal)? Are questions of human rights overridden by an Award on the inter-State sharing of waters?

In Shakespeare's *The Merchant of Venice*, Portia says that in terms of the contract Antonio could have his pound of Shylock's flesh, but without shedding a drop of blood. In the present case, she might have argued that the Gujarat Government could go ahead and raise the dam to 455 feet, but not displace more than 7,000 families (which was the number mentioned by the Tribunal). That number is evidently regarded as flexible: it has gone to upwards of 40,000 now; but the dam height of 455 feet is considered inflexible. In other words, the dam must be built to 455 feet, regardless of whether 7,000 or 40,000 or 1,00,000 families are displaced, and regardless of whether land for resettlement is available or not. Is that what 'non-reviewability' means?

The Madhya Pradesh Government had developed serious apprehensions about the feasibility of resettling and rehabilitating the large numbers of people involved and had proposed a reduction in the height of the dam to 436 feet to minimize displacement and make the task of rehabilitation more manageable. Others (scholars and analysts) had proposed alternatives that might have envisaged still lower heights and reduced displacement dramatically. Those propositions may or may not have survived a careful scrutiny. However, they were ruled out of consideration altogether by the simple argument of the sanctity of the Tribunal's Award, and, that argument has been accepted and endorsed in the Supreme Courts judgement. Let us suppose for a moment that Madhya Pradesh had proposed an *increase* in the height of the dam for enhancing power-generation: what would have been Gujarat's reaction? Again, in the hypothetical event of a generous Madhya Pradesh Government offering to reduce its share of Narmada waters and let Gujarat have 2 million acre-feet (MAF) more, would Gujarat have refused such an offer on the ground of the sanctity of the Tribunal's Award?

The details mentioned in the Tribunal's Report have not in fact been treated as immutable. The power-house configuration has been completely changed. Changes (liberalizations) have been made from time to time in the R&R policies and packages. The Supreme Court itself (in an earlier decision) have modified the time limit laid down by the Tribunal for the completion of rehabilitation arrangements (in relation to submergence of land). If all these changes were considered acceptable, then why should the suggestion of changes in the physical features of the project to minimize human suffering be considered improper and unacceptable?

Besides, the fact that the project was mandated by a Tribunal was not held to exempt it from the usual procedures of techno-economic examination and approval. The need for approval implied the possibility of non-approval. If the examination had resulted in a negative finding, or in an approval subject to some modifications, would that have constituted a violation of the Tribunal's Award? If new facts come to notice that show that a dam of the prescribed height at the stated place might be dangerous, would it be nevertheless obligatory to build it? This is not a hypothetical question. The recurring tremors in the Koyna area and the occurrence of the Latur earthquake (both in Maharashtra) seem to call into question our earlier understanding of the nature of seismic activity in central and southern India. Does this call for a review of the safety aspects of the Sardar Sarovar Project, or should we take the view (as even Justice Bharucha does) that those aspects have already been studied adequately? If in fact the hydrological assumptions of the Tribunal were wrong, would that call for a review of project design, or would the specifications laid down by the Tribunal preclude such a review? What absurdities we are driven to by the 'non-reviewability' argument of the Gujarat Government which has been upheld in the judgement!

Conclusion

In conclusion, the judgement was a negative answer to those who sought relief, and a severe blow to peoples' movements. It was received with dismay not only by the NBA but by many others across the country who have been concerned about the sufferings of the people affected by such projects. On the other hand, the judgement was received with jubilation by the Gujarat Government. It appeared very likely that the judgement would lead to a hardening of attitudes and to even greater intolerance towards dissenting opinion than before, and that the fragile façade of politeness maintained with difficulty earlier would now disappear. As for water resources planning in general, it could be expected that there would be even less receptivity than before to pleas for a reorientation and for a consideration of alternatives to big dams. One felt inclined to exclaim 'Cry, the Beloved Country'.

(Note: The apprehensions about a hardening of attitudes expressed in the last paragraph were amply borne out by the anger and intolerance with which the Report of the WCD was received shortly after the Supreme Courts judgement. This is discussed in Chapter 15.)

15

World Commission on Dams and India: Analysis of a Relationship

Introductory

The Report of the World Commission on Dams (WCD) entitled '*Dams and Development: A New Framework for Decision-Making*' was released in London in November 2000. Its findings on past global experience in relation to dams, as well as its recommendations (approaches, policies, procedures) for the future, were promptly and comprehensively rejected by the Government of India's Ministry of Water Resources (MoWR). The present chapter is intended not as a detailed examination of MoWRs' criticisms, but as an attempt to throw some light on the *nature* of their response. (The author, as a member of the team that had produced an India country study for the WCD, cannot claim to be a disinterested observer, but will try to present as fair and objective a picture as he can. This has been written in his personal capacity and not on behalf of the India study team.)

MoWR's reactions to the WCD Report were set forth briefly but categorically in their letter No.2/WCD/2001/DT (PR) Vol.-III dated 1-2-2001 addressed to the Secretary General of the WCD. A copy of this should be available on the WCD's Website (*www.dams.org*). Their detailed comments on the WCD Report were given in a long statement that seems to have been posted on the Internet later in the same month (see *http://genepi.louis-jean.com/cigb/inde.htm*). A third source from which we can get an idea of MoWR's thinking is the article on the WCD Report by B.N. Navalawala (then Adviser in the Planning Commission, and later Secretary, Ministry of Water Resources) published in the issue of the *Economic and Political Weekly* dated 24 March 2001. That article was doubtless written in the personal capacity of the author, but the criticisms that it makes are similar to (though expressed in more careful and restrained language than) MoWR's comments.

As mentioned, MoWR's response was comprehensively negative. They questioned the composition of the WCD, its procedures, the adequacy and representativeness of the sample studied, the 'knowledge base' behind the Report, the manner in which the Report was finalized, and the fairness and objectivity of the analysis and findings; and they found Part II, which outlined an approach to future planning and set forth criteria, principles and guidelines based on that approach, totally unacceptable. Moreover, their comments were expressed in unusually strong language. This was not mere non-acceptance but total denunciation.

Reasons for Harsh Response

What was the explanation for this extremely negative and harsh response? The answer is twofold. First, against the background of the prolonged and bitter battle over the Sardar Sarovar Project (SSP) and the increasing polarization of attitudes on the large-dam controversy, the Indian water resources establishment tended to react with dismay to the very idea of the setting up of a WCD, which seemed to them a sinister 'anti-dam' move. The Government of Gujarat tended to see this as a conspiracy against and a threat to the SSP, and their perceptions had a strong influence on the Government of India (GoI). (In particular, a reference to the 'decommissioning of dams' that figured among the many subjects for study by the Commission set alarm bells ringing: the suspicion was that this was an attempt to scuttle the SSP.) Secondly, the establishment's suspicions and hostility were particularly aroused by the membership of the Commission, which included Medha Patkar (whom they regarded as their arch-enemy) and L.C. Jain (whom they considered to be her friend and sympathizer). They were therefore, quite understandably, antagonistic to the WCD from the start.

That background conditioned the GoI's attitude to the WCD all along. They stayed out of the consultative 'WCD Forum' that was established. When the WCD began its work and wanted to hold a hearing in India early in 1998, the GoI was initially not averse to this, but subsequently, under pressure from the Gujarat Government, advised the WCD that the proposed hearing would be inopportune. (There were newspaper reports to the effect that the Gujarat Government even threatened to arrest the Members of the Commission if they were to enter the State!) Some time later, when the WCD commissioned a country study of India's experience in relation to large dams by a team of five persons, as well as a cross-check survey of five dams by another group, the GoI's hostility to the WCD was reflected in their general attitude to these groups, though individually some members of these teams were able to hold discussions with and get material from senior officials in the Central Water Commission (CWC) and the Planning Commission. At a very late stage, when the WCDs' work was well advanced, the GoI decided to join the consultative Forum as a member. MoWR and the CWC also participated in two 'Stakeholder Consultations' held by WCD at Chennai (in collaboration with the Madras Institute of Development Studies) and Delhi (in collaboration with the Indian Institute of Public Administration) in February–March 2000 for considering a draft of the India country study prepared by the team of five (as mentioned earlier). This phase of 'engagement', however, lasted only for a few months. MoWR/CWC did not like the country study, and perhaps this led them to feel that they should dissociate themselves from the WCD process. At any rate, by the time that the WCD came out with its own Report, the GoI had virtually dropped out of the Forum, and had reverted to its original coldness to the WCD. The final chapter of the story was the release of the WCD Report in November 2000, and MoWR's comprehensive rejection of that Report.

Criticisms of India Country Study

Let us now consider the nature of the criticisms voiced by MoWR. First, we take up MoWR's response to the India country study entitled '*Large Dams: India's Experience*' submitted by a team of five to the WCD in June 2000. The WCD sent a copy of this study to MoWR and their response was one of 'outright rejection'. Their criticisms were fourfold: (i) they questioned the competence of the members of the team; (ii) they felt that the study was unbalanced and negative; (iii) they criticized the fact that the Report itself reflected differences among the Members; and (iv) they charged the team with not revising their report in the light of the comments and information provided by MoWR and the CWC.

In so far as the question of competence is concerned, we need not go into the details of the actual terms used in MoWR's comments or in the *EPW* article, but may take the criticisms to mean that the members of the team were unsuitable for the task at hand. The readers can judge this for themselves in the light of the composition of the team, which was as follows: R. Rangachari, a distinguished engineer and former member of the CWC; Dr Nirmal Sengupta, a well-known professor at the Madras Institute of Development Studies and sometime consultant to the UNDP, FAO, and the Netherlands Minister for Development Cooperation, and author of several publications on water and irrigation (see bibliography); Shekhar Singh, environmentalist, former adviser at the Planning Commission, currently teaching at the Indian Institute of Public Administration (IIPA), a member of the Ministry of Environment and Forest's Environmental Appraisal Committee for river valley projects, as also of the Environment Sub-Group of the Narmada Control Authority for many years; Dr Pranab Banerji, economist at the IIPA; and the author. The question of competence will not be further discussed.

As regards the charge that comments and information provided were not taken into account, this is not true. As mentioned earlier, during and after the two 'stakeholder meetings' of February–March 2000 at Chennai and Delhi, MoWR/CWC provided detailed comments and information. Despite the severe time-constraint, all this was duly taken note of, and such revisions, corrections or additions as seemed necessary to the authors were carried out. There were also some discussions and clarifications in this process. Revisions were also made in response to comments and material received from numerous other sources. This exercise of correction/revision was quite a substantial one, and the final Report was submitted by the team to the WCD in June 2000. However, MoWR and the CWC were evidently dissatisfied with the results, as their considered response to the Report was one of 'outright rejection'. They would doubtless have been happier if the authors had rewritten the Report completely to conform to their (MoWR's/CWC's) views and perceptions. When this did not happen, they were disappointed. Their accusation of 'non-revision' really means that the revision was not to their satisfaction. In other words, this is not a case of 'non-revision' but one of persisting differences of perceptions and views. The study team cannot be faulted for this.

The charge of differences within the team and of inconsistencies is difficult to understand. In the preface as well as in Chapter 7 it had been clearly explained that the Report was not in its totality a joint effort; that the different chapters were written by different members; that each took responsibility only for his chapter(s), and was not necessarily in full agreement with everything in the chapters written by the others; and that all five had come together and taken joint responsibility for the final Chapter entitled 'Some Agreed Conclusions'. Where is the scope for misunderstanding or criticism in this? There were indeed some differences of perceptions and views among the members, but all of them did agree on the 'agreed conclusions'; and so far as one can see, there are no inconsistencies or contradictions within the agreed conclusions.

Coming to the criticism that the Report was negative and not balanced, this is really an expression of disagreement on the part of MoWR. What they mean is that they themselves would have put the balance between the positive and negative aspects of dams at a point different from the point at which the report (Chapter 7) appears to place it. They are entitled to their views. This point will not be debated here. Let the readers decide whether the Report presents a fair picture or not. Unfortunately, the Report has not been published and may not be easily accessible, but it can be looked up on the WCDs' Website.

World Commission on Dams: Composition

Turning now to the WCDs' own Report, we must remember how the Commission came to be set up. By the 1990s, the movement against dams—not only in India but elsewhere as well—had led to a strong polarization of opinions and to an *impasse* in respect of large projects. The World Bank had withdrawn from the SSP in India and Arun III Project in Nepal, and had become wary of new projects. At the operative level, it would have liked to get back into the business of lending for such projects, but there were difficulties at the level of the board; some of the country directors were unenthusiastic about big dams. The dam-building industry wanted to end the *impasse*, as did the equipment-supplying industry. Some governments too wanted to bring the prevailing uncertainty to an end. The 'Greens', on the other hand, and NGOs involved in resettlement and human rights campaigns, were opposed to dams. It was in an effort to find a way out of this logjam that the unprecedented Gland Consultation of April 1997 took place. All interests (dam-builders, equipment-suppliers, 'Green' and human rights NGOs, the World Bank, the International Union for the Conservation of Nature (IUCN) and so on) were present. (The International Commission on Large Dams ICOLD), a body of engineers, and the Government of China were among the participants.) It was at that meeting that the decision to establish an independent commission was taken. The independence of the Commission was sought to be ensured partly by distancing it from the World Bank, and partly by diversifying its sources of financing. The constitution of the Commission was the subject of prolonged negotiation with a view to ensuring acceptability to the diverse interests and points of view involved. The Commission, as finally set up, included representatives of all the interests and points of view (dam-builders, suppliers of equipment for dam-building, managers of river valley projects, environmentalists, opponents of dams, persons concerned with the interests of aboriginal communities and so on) that had been present at Gland. The composition of the Commission may be seen in the Report.

A question often asked in India, not only in official circles but elsewhere as well, is: 'Why should both the representatives from India be of the anti-dam persuasion?' The reference is to L.C. Jain and Medha Patkar. L.C. Jain will deny that he is 'anti-dam', but leaving that aside, two points must be noted. First, there was no country representation; the members were there in their individual capacities and not as representatives of their countries. They could indeed be said to be 'representative', but of points of view and not countries or institutions. Secondly, even assuming that both Medha Patkar and L.C. Jain were 'anti-dam', there were other members representing the 'pro-dam' point of view: the past chairman of the ICOLD, the CEO of Asea Brown Boveri (suppliers of equipment for dam-building), the Chief Executive of the Murray Darling Basin Organization of Australia, etc. The Commission as a whole could not be said to be either 'pro-dam' or 'anti-dam'. The Chairman was a very distinguished Minister in the South African Government (then Minister for Water Resources), and by no stretch of the imagination could he be described as an 'anti-dam' person. (It may be mentioned here that the Chairman, Prof. Kader Asmal, was later awarded the Stockholm Water Prize.)

Criticisms of Report

MoWR's comments on the Commission's procedures, the adequacy and representativeness of the sample studied, the knowledge base of the WCD, the manner in which the Report was finalized, its fairness and objectivity, etc., will not be dealt with here. (A restrained rejoinder was published by S. Parasuraman, a senior adviser in the WCD, in an article in *The Hindu* on 11 September 2001).

Contributions of Dams: Questions

However, one point needs to be put in proper perspective. One of the criticisms is that the positive contributions of dams have not been sufficiently brought out. The WCD could argue that it (perhaps unwisely) took this for granted. It is of course true that dams provide irrigation, electricity, etc. Why else would dams be built? But five questions would arise in each case (leaving aside the purely managerial question of time and cost overruns):

(i) Were the expected benefits fully realized? Or were there serious shortfalls?

(ii) At what cost (financial, environmental, social, human) were those benefits achieved? Was the final balance between the costs and the benefits (as actually achieved) positive or negative?

(iii) Going beyond the cost-benefit calculus, were there serious adverse impacts and consequences (environmental, social, human)? Were all of them remediable? Were all of them foreseeable? Taking everything into account, was it right to have implemented the project in question?

(iv) What have been the equity implications? Has the project (during the planning, construction, implementation and operation stages) been a *positive* contributor to equity and social justice, or a *negative* contributor, or was its impact neutral?

(v) Assuming that certain benefits were aimed at and achieved, was the project the only way of achieving those benefits? Were there alternative answers that might have avoided some of the negative impacts? Were such alternatives and options considered?

It will be seen therefore that a mere reference to irrigation, hydro-electric power, etc., will not take us very far. To claim these as the contributions of dams is to beg the questions listed above. Such a claim is made easily enough and often enough; it is all the other (negative) points that need to be looked at carefully in a critical examination of dams. In the nature of things, such an examination is bound to appear negative. (Incidentally, the determination of the 'contribution' itself is sometimes problematic: should we attribute 'food security' to *dams* or to *irrigation*? How much of the irrigation comes from dams? Could that irrigation have been ensured through other means? These questions cannot be easily answered, but they are important and need to be asked. This too would appear 'negative'.)

Is the Report Anti-Dam?

Assuming that the Report could have said something more about the positive contributions of dams, is it open to the charge of being 'anti-dam'? There seems to be no basis for such an accusation. Nowhere does the Report say that dams should not be built. It assumes that dams will continue to be built, and the whole of Part II of the Report is an elaborate prescription of how dams should be planned, approved, built and operated in the future. In fact, those of an anti-dam persuasion might well complain that the Report does not really squarely face the question implied in its first Term of Reference, namely, whether, on balance, dams are good or bad. Implicitly it accepts the need for dams. Its criticism is not against dams themselves, but against the manner in which projects have been planned, evaluated, approved and implemented in the past. Its recommendations in Part II are therefore directed towards better planning and decision-making in the future. However, assuming that all past deficiencies in this regard are eliminated, that 'people' are involved in planning and implementation from the beginning, that corruption and collusion have disappeared, and that equity and social justice have become central concerns, can we take it that large dams will be wholly good? For instance, will they be environmentally benign and conducive to the sustainability of planet Earth? That crucial question is not really posed and answered in the Report. How then can the Report be described as 'anti-dam'? (Incidentally, to repeat a point made in an earlier chapter, the very use of the term 'anti-dam' as a term of abuse is revealing. Why is it right to be 'pro-dam' and wrong to be 'anti-dam'? Are not both partial perceptions?)

Negative Perceptions of Governments

Be that as it may, the sad fact is that the official circles and the engineering community in this country regard the WCD Report as being essentially anti-dam, and this tends to have some influence even beyond those circles. It must be added that the GoI's negative perceptions of the Report are partly shared by several other Governments in this part of the world. In February 2001 the Asian Development Bank held a meeting at Manila to discuss the Report. (The author was an invitee.) The GoI did not participate in that meeting (as was only to be expected), but other governments of the region were well represented. The leader of the Chinese team expressed himself very strongly and made points similar to those made by the Indian MoWR in their posting on the Internet. The representatives of other Governments, namely, Pakistan, Bangladesh, Nepal, Sri Lanka, the Philippines, Vietnam, etc., were more restrained in their language, and were prepared to find some merit in the Report, but in essence they were all against the approach and pro-cedures suggested in the Report. All these countries had dam-building programmes, and felt that their efforts might be hampered by the Report if it were to gain general acceptance. (It must be noted that while the voices of the dam-building Governments were prominent—that of China was particularly strident—other points of view were not much heard at the meeting; the voices of NGOs were few and muted.)

Western Conspiracy?

An important strand in the reactions of developing countries to the Report is to see it as yet another instance of the imposition on them by the developed countries of an agenda designed in the latters' interests. There are indeed deep and justifiable concerns on the part of developing countries regarding the dubious use that the developed world tries to make of the World Trade Organization forum and the processes of environmental negotiations such as those relating to greenhouse gas emissions. What the developed countries have been trying to do is to impose severe disciplines on the developing countries without making the slightest changes in their own resource-intensive and 'toxic' ways of living (to use Dr Anil Agarwal's word), which have brought the world to its present state, and which continue to exercise a baneful influence on the rest of the world by providing a definition of the idea of devel-opment. This is quite rightly resented by the developing countries. The author shares these concerns. Unfortunately, that resentment tends to distort thinking and to get carried into inappropriate contexts, such as the controversy over large dams. Criticisms of the adverse consequences of dams, and the recommendation of caution in taking up such projects, are also perceived as a Western conspiracy to prevent poorer countries from developing. There is the argument, often repeated, that the West, having built all the dams that it needed to build, is trying to prevent

India from building any. Such perceptions, often accompanied by strong nationalistic sentiments, are difficult to argue against.

Campaign against WCDs' Report

In fact, what we have here is not a reasoned argument. The attempt is not to *deal* with the Report, but to discredit it so that it does not have to be taken seriously. The dam-builders seem to have decided to undertake a campaign towards this objective. Lest this should seem a fanciful hypothesis or a conspiracy theory of the author's own, it may be mentioned that a letter from the Indian Committee of the International Hydropower Association signed by the Executive Chairman and President, ICOLD, inviting the addressees to a conference in February 2002 gives clear evidence of a concerted effort to render the WCD Report ineffective (letter no. 395/CPU/IHA dated 22 March 2001, not reproduced in full here). It claims the following 'achievements':

1. For the first time the three organizations—International Commission on Irrigation and Drainage (ICID), International Hydropower Association (IHA) and International Commission on Large Dams have come on a single platform and these represent 25,000 members in 81 countries.
2. Government of India have issued official rejection of the report of WCD.
3. We were able to get the WCD meeting scheduled in India for 12 February 2001, basically for propaganda purpose, cancelled.
4. Acceptance of the WCD Report by funding agencies, specially the World Bank, scheduled on 15/16 February, 2001, did not succeed in spite of the high profile of the WCD.
5. Similarly, we could interact with other countries specially China who could also influence Asian Development Bank not to accept the WCD report.
6. Finally at the Cape Town meeting in the last week of February 2001 considerable amount of pressure could be exerted by Forum members so as not to accept the WCD report in toto.
7. The various industries connected with power development in the developed countries, especially in the USA, were advised so that they could join hands with the developing countries to oppose the acceptance of the WCD Report in their own interest.

8. We have been able to address the President of the OECD not to accept the WCD report in their meeting in Paris.

The motivations behind this campaign are not being questioned: they may be entirely honourable. Quite possibly, those who are interested in building dams (governments, international organizations related to dam-building) are convinced that the world needs many more dams; that this constitutes 'development' and is for the good of humanity; that the WCDs' Report is likely to come in the way of this noble enterprise; and that it is therefore necessary to eliminate that danger. It is, however, clear that we are no longer in the realm of civil discourse; this is war. This explains the ferocity of the attack on the WCD (perceived as the enemy) and the attempts to denigrate anyone associated with it in any way.

(Incidentally, dams seem to have become the 'temples of modern India' in a manner that perhaps Nehru never dreamt of when he used that phrase. They seem to inspire a fervour akin to religious emotion; and the perception of a (real or imagined) threat to the building of more such temples calls forth a *furor religiosus*.)

Negative Reactions to Accepted Principles and Procedures

It may be argued that the WCDs' Report is in effect 'anti-dam' because the principles and procedures recommended in Part II are so stringent and cumbersome as to make it very difficult for any project to go through in the future. It is this reaction that is the greatest cause for concern. The points about cumbersome and dilatory procedures and rigid insti-tutionalization need to be considered. Adaptations and modifications can and must be made to the WCD guidelines from a practical point of view, with due regard to the circumstances of different countries and cases. What is worrying is the negative reaction to unexceptionable principles and approaches.

Broadly speaking, what does the Report advocate? It wants to improve planning and decision-making in the future by drawing lessons from past experience. It restresses the Dublin–Rio principle of 'stakeholder consul-tation and participation' (we shall revert to that term later) and recom-mends the requirement of 'free, prior, informed consent'. It formulates a 'rights and risks' approach that transcends the old-style cost-benefit analysis. (The reference is to the impact of a project on the rights of affected groups and the risks that they have to bear.) It underscores the imperative of 'sustainability'. It urges the consideration of options and alternatives. How can anyone disagree with any of this? And yet, the Governments and international bodies concerned with dam building felt so threatened that they considered it necessary to oppose the WCDs' Report fiercely.

Not very long ago, even project planners and dam-builders would have been willing to agree that dams had had some adverse consequences, that past processes of planning and evaluation had been deficient, that environmental and human aspects had not received adequate attention in the past, that project-affected persons should be consulted at a much earlier stage and should be given some rights over the benefits arising from the project, and that significant changes in attitudes, processes and procedures were called for. Observations and recommendations on these lines will be found in numerous official documents and reports. Unfortunately, the perception of the Report as a threat to national planning and policy-making led to a closing of ranks, a hardening of attitudes, and a retrogression from the degree of enlightenment that was beginning to emerge. There was a downgrading of environmental, social and human concerns.

Intolerance

The spirit of intolerance is in the air. It was evident in the hostile and uncivil interventions by some of the MoWR/CWC representatives in the two stakeholder meetings held at Chennai and Delhi in February–March 2000 (noticed by all present, including members of the media), as also in several other subsequent meetings and conferences relating to water resource development. References to the adverse impacts of dams are ill received. The espousal of environmental concerns, equity and social justice by individuals and NGOs is misinterpreted. Genuine and profound concerns are made to appear sinister, dishonourable and anti-national. Terms such as 'environmental lobby', 'eco-fundamentalism', 'enemies of development', 'foreign agents', etc., are freely bandied about. There is a strident reassertion of the dominance of the engineering point of view, and a deprecation of other perspectives.

Disturbing Observations

In MoWR's comments on the recommendations in Part II of the WCD's Report there are some disturbing observations. Consider the following:

(i) On equity: 'However, emphasis on equity in a wrong manner is dangerous. Many countries including India and USSR have learnt the hard way (*sic.*) that too much emphasis on equity can only perpetuate poverty. ...'

(ii) On the consideration of alternatives and options: 'Only developed countries, which have the time and money to explore all possible alternatives to dams can afford, if they wish, may wish (*sic.*) to opt for such exercises on 'Options Assessment' as brought out in WCD's Report.'

(iii) On the principle of consultation and participation: 'If for every single project decision, informed participation and acceptance by all groups is to be carried out, the decision-making would become a long drawn, protracted process ...'; '"Free, prior and informed consent" as suggested by the Commission is likely to render all major project proposals of significance subject to purely local perspective and evaluation, negating the regional and national planning of economic development.'; 'Tribals and even non-tribals affected by the project would understandably view the dam proposal from their own perspective. ...'

A Retrogression

It is not the intention of this chapter to suggest that MoWR's detailed critique of the WCDs' Report is without merit. There are many points in it that need careful responses. In essence the argument is that the developmental needs of developing countries necessitate dam projects; that the WCD has ignored these needs; that water-scarce areas require water-transfers from elsewhere to supplement local resources; that while considering the interests of people who are affected adversely by such projects, we must not forget the interests of those in the command area who will receive irrigation or drinking water from the project; that our concern about the possible environmental impact of a dam in certain areas should not lead us to forget the beneficial environmental/ecological impact of the transfer of water to water-short areas, and so on. It is the author's view that there are some fallacies here, and that the arguments are not as strong and persuasive as MoWR might think; but there is hardly any doubt that the points are important and demand detailed discussion. However, what is regrettable is that the anger of the dam-builders has led them to rebel against even accepted principles such as equity, consultation, participation, consideration of options and alternatives, sustainability, etc. It is sad to see that having taken several steps forward during the past decade or two, MoWR (and others of their way of thinking) have now taken at least some (if not many) steps backward.

The Notion of 'Stakeholders'

While not entering into a detailed discussion on the many issues raised by MoWR, it seems necessary to discuss at least two of them. The first is regarding the notion of 'stakeholders'. Consider this passage from the comments made by the MoWR:

'While repeatedly talking about stakeholders, those sections of society which have a strong stake in a dam construction and who stand to suffer and lose if a dam is not constructed or is delayed, are not even recognised as stake holders!'

As pointed out earlier, the concept of 'stakeholder' is a flawed one that has great potential for misuse (see Chapter 8, section II). We need not go over that ground again here, but must repeat the point that the vital difference between project-affected people and prospective beneficiaries tends to get blurred by the bland assimilating term 'stakeholders'. There is a cruel irony in describing the involuntary and helpless victims of a project as 'stakeholders', and this is compounded when they are put on the same footing as those who stand to benefit from the project. The primacy of the former over the latter needs to be recognised.

State and Civil Society

The second point that needs to be noticed is the difficult question of the relationship between the state and civil society. MoWR adopts the simple position that in a parliamentary democracy such as India people elect their representatives to *panchayats*, legislative assemblies and Parliament; and that their interests can be presumed ᵗo be taken care of by these representatives, and by the executive governments that are accountable to them. This view would call into question the need to 'consult' people separately in the context of a project; and yet many democracies have found it necessary to institute consultation procedures. Further, in such a view, there would be no question of any conflict between the state and the people; the distinction between state and civil society would be pointless; and there would be no need at all for any movement for 'empowering' the people. It is clear that the point of view set forth by MoWR has many serious implications. The debate will not be pursued here; it is merely brought to the reader's notice for being pondered over.

Is the Ministry of Water Resources Speaking for Government of India as a Whole?

Finally, the matters discussed in the WCDs' Report and in MoWRs' comments go far beyond the domain of that Ministry. There are other Ministries charged with the responsibility for agriculture, watershed development, rural development, rural and urban water supply and sanitation, environmental concerns, public health aspects, the interests of project-affected persons who often belong to scheduled castes or tribes or other backward classes or weaker sections of society, the welfare of women and children and so on. It seems very difficult to believe that all those Ministries would readily subscribe to some of the statements made in MoWRs' comments. Were there any inter-ministerial consultations before MoWRs' comments were communicated? Are the various Ministries concerned willing to endorse everything that MoWR has said? Would that be in conformity with the policies of the respective Ministries? What does the Planning Commission think about these matters? Do the comments represent the views of one Ministry or those of the GoI as a whole?

A Threefold Plea

As mentioned more than once, this is not a defence of the WCD or a rejoinder to MoWRs' points. That is a debate that needs to be separately pursued. What this article has been trying to show is that the anger of the dam-builders has tended to prevent such a debate; instead, an effort has been mounted to deflect a perceived threat by denigrating the Commission and those who had made contributions to the WCD process and damaging their credibility. This chapter will conclude with an earnest threefold plea:

(i) to MoWR/CWC, their counterparts in the state governments, their international *confrères* (ICOLD, ICID, IHA, etc.,) and the engineering community in general: *call off hostilities and return to the path of rationality and civil discourse*;

(ii) to the Ministries (and their counterparts at the State level) concerned with the environment and forests, women's and children's welfare, scheduled castes and tribes, agriculture, rural and urban development, community initiatives in water-harvesting and watershed development, etc.: *ask yourselves whether your statutory responsibilities and your departmental concerns, policies and guidelines are adequately reflected in the MoWR/CWC response to the WCDs' Report, and further, whether perhaps that Report contains ideas that have some resonance from your points of view* (this is also addressed to the Planning Commission); and

(iii) to the media, academics of diverse disciplines, journalists, doctors, teachers and other professionals, the intelligentsia in general and indeed the ordinary citizen: *take a modest degree of interest in this important debate in which major issues and concerns are involved, and do not ignore this as a debate between two sets of specialists.*

It is hoped that this threefold plea will not fall on deaf ears.

(Note: In recent months, there have been some faint signs of a recognition of the need for a debate, though the prejudice against the WCD continues to be strong. The reluctant beginnings of a debate stem less from newfound reasonableness than from a realization that the Report will not go away and needs to be dealt with. Be that as it may, the slight—though very slight—change in the atmosphere must be welcomed.)

16

Changing Views: A Personal Note

In recent years, I have been often asked about the putative change in my views on large dams. The question has been asked in diverse ways: with genuine and sympathetic curiosity; with sadness at my decline into error; with the intention of causing discomfiture; with anger and dismay at the perceived defection by an Establishment figure; and with gratification at the belated dawning of wisdom on a former bureaucrat. As the question of change is so often flung at me, accompanied by varying degrees of misunderstanding and misrepresentation, I thought that I should explain the position carefully and definitively to those who are interested. Those who are not are requested to proceed forthwith to the next chapter.

I am aware that in some quarters I am regarded as being anti-dam, anti-Establishment and anti-engineer. It would be tedious and pointless to undertake a refutation of such misperceptions. Instead, let me explain the evolution in my thinking in positive terms.

In the years 1985–87 when I was Secretary, Ministry of Water Resources, in the Government of India, there was no strong anti-dam movement. A degree of uneasiness had begun to be felt at the environmental impacts of such projects and the problems of displacement and rehabilitation, but very few at that stage argued against dams in general: the plea was for good EIAs or 'environmental impact assessments'. That was the climate of opinion, and my thinking was in conformity with it. However, I became convinced at an early stage that it was necessary to bring about a reorientation of thinking in the Ministry from an excessive preoccupation with large irrigation projects towards a wider and deeper engagement with issues of resource policy. That was the effort behind the drafting of the National Water Policy 1987, though the resulting document was an inadequate and imperfect one that showed the continuing influence of older ways of thinking. Another correction that I tried unsuccessfully to bring about was a move from an examination of a given and single project submitted for approval to a consideration of options and alternatives. Having said that, I must acknowledge that I accepted the need for a number of large projects, and piloted several through the machinery of government for approval. The Narmada Projects (Sardar Sarovar, Narmada Sagar) were among these. I played a role in the granting of 'conditional clearance' to these two projects early in 1987.

Soon after that the opposition to these projects began to gather strength, and in a couple of years, became formidable and acquired international dimensions. I pondered over the issue carefully and wrote a detailed article entitled 'Large Dams: The Right Perspective'. This was published in the issue of *Economic and Political Weekly* (EPW) dated 30 September 1989. The article aimed at comprehensiveness, balance and definitiveness. It was widely read and much appreciated by both the supporters and the opponents of dams. (It even had the distinction of being plagiarized in a theme paper in one of the National Water Conventions.) If there were a ten-point scale with position 1 representing a totally pro-dam position and 10 a totally anti-dam view, that article could have been put at point 5 in the scale, i.e., right at the middle.

Almost a decade later, I wrote another article entitled 'Water Resource Planning: Changing Perspectives', which was published in the *EPW* issue dated 12 December 1998. In operative terms, both the 1989 article and the 1998 one called for a consideration of options and alternatives, minimal recourse to dams (regarding them as projects of the last resort), stringent scrutiny and approval procedures and so on, but on the question of the overall balance of the costs and benefits of large dams there had indeed been a significant shift in my position in the aforesaid scale: from point 5 in 1989 to point 7 or 8 in 1998.

What brought about this change? Several factors: increasing awareness of the kinds of impact that such projects had and the difficulties of countering them; a better understanding of the limitations of EIAs; exposure to difficult and critical questions in seminars, committees and private discussions; and so on. An exchange between me and Jasveen Jairath in the pages of the *EPW* in 1989–90 (though at that stage I defended the 'balanced' position adopted in the 1989 article) kept working insidiously in my mind. (The exchange has been reproduced in Dhawan (1990). There was also an exchange of letters between me and Mihir Shah.)

However, it was my membership of two project-related Committees that caused a significant shift in my position: (i) the Five Member Group that went into various issues raised by the Narmada Bachao Andolan in relation to the Sardar Sarovar Project and submitted two Reports, the first (a unanimous one) in April 1994 to the Government of India, and the second (with one Member less, and a divided report) to the Supreme Court in April 1995 in response to four specific issues referred by the Supreme Court to the Group; and (ii) the Expert Committee on the Environmental and Rehabilitation Aspects of the Tehri Hydro-Electric Project. The material presented and the evidence tendered to these Committees, and the discussions that I had with the other members as well as with the wide range of persons who appeared before us, brought home to me the gravity of the environmental and human impacts of such projects and the enormous difficulty of countering them.

My growing doubts about the balance between the costs and benefits of such projects and about the soundness and adequacy of cost-benefit analyses as the basis for decision-making were deepened by my reading of Patrick McCully's *Silenced Rivers*, which is perhaps the strongest and best-documented statement of the case against large dams. (In a review of the book in the journal *Himal* I described the author as a prosecutor, and asked the defence counsel to step forward, but I have not seen any defence in the same class as McCully's critique.)

In 1998 I had to make a 'presentation' at the first public hearing held by the World Commission on Dams (WCD) at Colombo, and that gave me an opportunity to review my thinking on the subject. What I presented was a careful statement of both sides of the case; I described it as an unresolved controversy. My education and exploratory thinking on the subject were carried further by my work, as a member of a team of five, on an India country study on large dams for the WCD resulting in a Report entitled Large Dams: India's Experience in June 2000.

Subsequently, in 2001, a request for a paper on the large-dams controversy by a couple of American academics (for inclusion in a book on dams which they were editing in a series of volumes on various controversies entitled *History in Dispute*) provided me with an opportunity to make a considered statement on the issue. That contribution forms the basis of Chapter 11 in this book on the large-dams controversy. The same position is also reflected in Chapter 25 entitled "The Dilemmas of 'Water Resource Development'," and it is implicit in the chapters on the Narmada judgement and on the WCD. It is a closely argued, non-dogmatic, nuanced position. I refrain from re-stating it here. The present note is merely intended to explain whether, and if so, to what extent my thinking on the subject has changed over the years.

IV

The Language of Security

17

Scarce Natural Resources and the Language of Security

The Environment/Security Thesis

In recent years, the possibility of conflicts arising over scarce natural resources has evoked much interest not only in academia but also among 'think tanks', donor agencies and institutions devoted to strategic or security studies. Many seminars, conferences and workshops have been held on this theme. The purpose of the present chapter, which is based on a presentation made by the author in one such conference, is to draw attention to the implications of a certain proposition which has gained considerable currency.

The thesis, broadly stated, is that natural resource scarcities (or 'environmental scarcities' to use a term favoured by some scholars) are likely to lead to conflicts and violence; that these have security implications; and that these implications could translate into regional security problems. There are three key terms here, namely, 'environmental', 'security' and 'regional'. There are questions that need to be raised about all of them. However, before entering into that discussion, it may be useful to try and restate the thesis in the form of a narrower and more limited proposition, or couple of propositions, which avoid those problematic terms: (i) that the pressure on some vital natural resources is likely to become more and more severe; and (ii) that competing claims over scarce natural resources could lead to tensions and conflicts both within countries and between countries. There can be no disagreement with these propositions.

Taking water in particular as a good example of a scarce natural resource, we can think of many instances of conflict. Disputes between communities or countries over watercourses which cut across boundaries have an ancient history. Legend has it that the Buddha intervened in a terrible war between two communities in ancient India, the Sakiyas and Koliyas, over the sharing of the river Rohini, brought both sides together and ended the long-drawn discord. In modern times, there have been difficulties between Israel and its neighbouring countries, as also between Israel and the Palestinians, over water; disputes over the Tigris and the Euphrates; problems of water-sharing on the Nile between Egypt and the upper riparians, namely, Sudan and Ethiopia; problems over common rivers between the USA and Canada, and the USA and Mexico; in our own part of the world, protracted and contentious talks between India and Bangladesh over the Ganga (now resolved through a Treaty), and a long history of mistrust and suspicion between Nepal and India over projects on rivers (with the Mahakali Treaty marking — one hopes — a change); and within India, a bitter dispute between Tamil Nadu and Karnataka over the sharing of the Cauvery waters, which erupted briefly into violence in 1992.

(Incidentally, the Cauvery dispute and the Ganga waters dispute are only partly cases of scarcity of natural resources leading to conflict; more importantly, they are cases of political conflict leading to difficulties in resolving issues of resource-sharing, which, but for the politicization, might not have proved so intractable).

Both the problem of a scarcity of water (a scarcity which is already a reality in some parts of the world and which threatens to become a crisis in the not too distant future) and the likelihood of the scarcity leading to conflicts (again, a present reality and not merely a future prospect) are beyond question. However, difficulties arise when we enlarge the scope of the discourse and proceed to talk not about water or rivers but about 'environmental change' or 'environmental scarcity'; not about 'conflict' but about 'security', and not simply about 'security' but about 'regional security'.

'Environmental Change'

Taking the term 'environmental change' first, the desire to widen the scope of discussion beyond water or other specific natural resources is understandable. There are a number of ways in which acts of omission or commission in one area or one country can have impacts in another area or country; the configuration of groundwater aquifers may be such that exploitation in one country leads to depletion in another; one country may succeed in polluting the river waters that run to another;

deforestation in a mountainous country could lead to excessive flooding in a country in the plains, and the floods may also carry a heavy load of silt; industrial pollution in one country may affect the environment in another; if a dam were to burst in an upstream country there could be devastation in a downstream country. There could also be conflicts over shared seas or lakes. Acute scarcities in one country could result in large-scale migration to a neighbouring country, introducing strains in their relationship. The point need not be laboured further. It is clear enough that what one country does or fails to do in relation to natural resources and the natural environment in general can have serious consequences in other countries. These are matters for comprehensive and stringent regulations and clearance procedures, as also for inter-country understandings.

Whether a divergence of interests between countries in these matters would necessarily lead to conflicts or wars is debatable. We shall return to that question. Meanwhile it must be noted that all these are *limited* instances of 'environmental change'. There is a much larger sense in which the term has portentous implications; and that is when we consider the implications of global environmental change for the future of humankind.

When we reflect on matters such as the alarming build-up of carbon dioxide in the atmosphere, global warming, rise in ocean levels, drastic changes in weather patterns, the depletion of the ozone layer, the threat posed to aquatic life and coastal areas by oil spills from tankers (which are inescapable concomitants of the global trade in oil), the destruction of wildlife habitats and the extinction of growing numbers of species, the destruction of forests all over the world, the increasing likelihood of horrendous industrial and nuclear accidents (e.g., Bhopal, Chernobyl), the problem of disposal of nuclear waste, the exploding human population and the strains that this imposes on the carrying capacity of this planet, and so on, it is difficult to avoid apocalyptic visions of doom. Concerns like these led to the Rio Conference and to Agenda 21. That was an imperfect and flawed agenda, but even with reference to that Agenda very little has been done, as was evident during the deliberations of the First Assembly of the Global Environment Facility (Delhi, 1–3 April 1998).

Two points need to be noted in this context. The first is that these are *global* and not *regional* issues. The second is that the ways of living of the affluent West cast an irresistible spell on the rest of the world, and that it is this that accounts for the dangers which humanity and this planet face today. It follows from this that the essential conflict in relation to environmental change is between the rich 'advanced' countries and the poorer 'developing' countries. The primary responsibility for pulling humanity and planet Earth back from the brink of disaster must lie squarely on the former: unless they shoulder the burden of remedial action for their past depredations on nature, and are prepared to make drastic adjustments to bring down their 'standards of living' to sustainable levels, there is no hope of persuading other countries to accept environmental discipline or settle for modest ways of living which might seem akin to poverty by current Western standards.

Local conflicts over scarce resources, or over environmental concerns in particular regions, constitute one kind of problem; general and grave environmental degradation and the consequent dangers to planet Earth itself represent a very different set of problems. We are in danger of obscuring that difference if we use the same term ('environmental change') to describe both kinds of cases. Whether some in the West (politicians, big business) deliberately wish to blur that distinction and shift the focus of attention away from the West, is a question that cannot be gone into here; what we need to note is that when academics write or speak or hold conferences on such themes, they should be wary of unwittingly supporting obfuscatory efforts.

Conflict and 'Security'

Let us now turn to the term 'security'. Why should this term be introduced in the context of conflicts over natural resources or environmental concerns? What lies behind the transition from 'conflict' to 'security'?

Before attempting to answer that question it is necessary to be clear about the multiple senses of the word 'security'. There are the familiar security concerns of the Defence Ministries, military establishments, external and internal intelligence agencies and the internal police and paramilitary forces. Their worries include the capabilities of the armed forces of other countries and their weaponry and technology; the dangers of espionage, subversion, terrorism and destabilization; internal armed insurgency, riots and civil wars, and so on. We also use the term 'security' in the context of arrangements to look after the safety of industrial establishments, offices, shops, homes or individuals. In all these uses of the word, there are two components in the meaning: risk and protection.

In a different usage we talk about 'social security', 'food security', 'economic security' and so on; in these uses, too, the two components of risk and protection are present, but they are of a non-military kind. 'Social security' protects people against the risks and costs incidental to unemployment, old-age, ill-health, and so on; 'food security' means the assurance of an adequate availability of food for a growing population; and 'economic security' could mean measures to ensure that a country's economic system is not vulnerable to the risks of an external or internal debt-trap, sudden and large outflows of capital, a steep fall in the value of the currency, the acquisition by external agencies of excessive economic (and perhaps political) power within the country, and so on. Acute food or water scarcity could possibly result in internal instability, and in this sense may represent a threat to security in the conventional sense. Under certain circumstances a hostile country or agency may try to use means such as economic sanctions, cutting off essential supplies, placing an embargo on trade, etc., for coercive purposes, in which case the non-military and military senses of 'security' may tend to merge. However, it is useful to maintain the distinction between the two senses.

The question was posed earlier: What lies behind the transition from 'conflict' to 'security'? There could be many reasons: a desire to widen the horizon of the concerns and interests of the military establishment and to sensitize them to newer forms of conflicts and tensions, such as ethnic, cultural, natural-resource-related and environmental; a feeling that too narrow a preoccupation with security in the conventional sense may not enable them to understand inter-country relations and politics in all their complexities and may limit their capabilities even in the security area; and, of course, the apprehension that some of these non-military disputes and tensions may have the potential of turning into hostile relations or even wars. No one can seriously oppose the widening of the mental horizons of the military establishments or with attempts to sensitize them to unfamiliar kinds of tensions and dissensions. One can even consider the hypothetical possibility of utilizing the special skills or capabilities available in the military establishment in the efforts to resolve conflicts in the sphere of natural resources or environmental concerns (though it is not easy to imagine what those skills and capabilities could be). What causes concern is the idea that such conflicts have security implications in the conventional sense.

'Water Wars'?

At this juncture it is necessary to take note of the currently fashionable thesis that future wars may be fought not over oil but over water. Many experts have been saying for some time that as supplies of fresh water in nature are finite, and as the world population has increased and is continuing to increase to staggering magnitudes, humanity is going to face severe water scarcities in the next few decades, and that the severity of the water stress will be very great in certain regions of the world. There are no grounds for questioning that prognosis. However, it does not follow that this will necessarily lead to water wars.

First, the analogy with oil is misleading. A powerful country can go to war with a weaker one over the latter's oil resources, acquire control over those resources, exploit them commercially through the establishment of corporations and convert them into products for export, or transport the oil itself through pipelines, tankers or bulk-carriers to distant parts of the world. None of this applies to water. There is no comparable international trade in bulk in water (apart from trade in bottled drinking water). The water resources of one country can at best be used in the neighbouring countries, and not piped or transported by tankers or vessels to countries at enormous distances. One cannot look far into the future, but it is difficult to visualize a vast international trade in water. In any case, supplies are finite, and may not be able to sustain such trade; few countries are likely to have large exportable surpluses of water. The largest use of water is in irrigation and it is difficult to imagine a water-short country importing water in bulk for irrigation from a distant source. Water resources are likely to be harnessed essentially for local use, 'local' here meaning one country or a few countries adjoining one another.

Secondly, given that situation, those countries are much more likely to cooperate with one another and enter into agreements or treaties, or embark on joint projects for water resource development and utilization, than to go to war. Here again, the analogy with oil breaks down. It may be feasible to conquer a country and take way its oil; it may not be so easy to take away a country's water. It is only under very special sets of geo-political circumstances that a country can hope to acquire control over head-waters or aquifers in a neighbouring country through military means. Even in such a case it may be simpler for the country needing water to enter into an agreement with the country which has the water resources. A war over water may be theoretically possible, but it seems unlikely to happen: not in Asia, and perhaps not even in the Middle East.

In fact there are large numbers of examples of treaties, agreements, joint commissions and so on, over water. In 1996 India signed treaties with Nepal over the Mahakali river and with Bangladesh over the Ganges; much earlier in 1960 India and Pakistan entered into a Treaty on the Indus system. It is remarkable that despite the extremely difficult relationship between India and Pakistan that Treaty has continued to operate and has not been repudiated, even during the three wars between the two countries. India and Pakistan have gone to war several times over other issues but not once over water. It was in fact through the deliberate isolation of this particular matter from security concerns and establishments that the Treaty has survived despite wars.

There are also other examples of inter-country cooperation such as the Mekong Commission, the Nile Commission and so on. The evidence is overwhelmingly in favour of the thesis that countries will enter into understandings or agreements over water, and not in favour of the proposition that they will go to war. Indeed, not merely is there a possibility of avoidance of conflicts or resolution of conflicts, but there are even prospects of positive and pro-active cooperation over natural resources and environmental concerns in general.

As population grows and water-stress increases, two things may happen: first, there may be an increased number of agreements or treaties between neighbouring countries or among a group of countries in a region, and joint commissions or other institutions may be established; secondly, people may learn to live with a reduced availability of water and try to get more and more value out of a limited quantity of water through improvements in water-use efficiency and technological innovations. Maximizing the value per unit of water may become an important area of research.

Dangers of 'Security' Language

The thrust of the above argument is not to deny the possibility of conflict over scarce resources or over environmental impacts. We must certainly recognize that possibility. However, from that point on we can take one of two alternative routes: we can proceed on the path of conflict-resolution, harmony and cooperation; or we can talk about 'security implications' and visualize the possibility of war. The latter approach is unlikely to promote harmony or cooperation. The postulation of a security angle to such conflicts may become a self-fulfilling prophecy.

Military and intelligence establishments and defence analysts must of course necessarily examine all possible foci of conflicts and tensions from their own specialized and professional points of view; but it is hardly necessary for academic and research institutions dealing with water and other natural resources, or with environmental issues, to adopt the 'security' language, or allow themselves to be co-opted by the military and security communities. The danger of bringing environmental or natural-resource-related conflicts under the 'security' rubric is that it tends to induct the wrong personnel, namely, military and intelligence experts, into this area; and to subsume such conflicts under a major (essentially adversarial) category which may enhance their negative potential, render them more intractable and perhaps even incorporate them as components in military or strategic planning.

Finally, reverting to the argument of the previous section, 'environmental scarcities' are the result of a flawed relationship with nature and distorted notions of 'development' and 'civilization'. We can focus on the conflicts that such 'scarcities' generate and the 'security' implications of such conflicts; or we can focus on the sickness that causes the conflicts, and try to remove that sickness through a global cooperative endeavour. Harmony between nations is not possible without harmony between humanity and nature. The language of 'security' will never enable us to realize this.

Let us assume for a moment that there is a cluster of nations with no conflicts at present among them, all of them together using a finite natural resource (which at the moment seems relatively abundant) at an excessive rate. If there is no conflict there is presumably no security angle, but disaster is surely in store; and that disaster will be averted not by injecting a security consciousness (related to the possibility of future conflicts) into the situation, but by promoting the idea of cooperation in a more healthy and sustainable use of the resource in question.

The 'Regional' Angle

The 'security' slant becomes even more dangerous when it is qualified by the word 'regional'. The whole idea of 'regional trouble spots' is a post-

Cold War formulation essentially from the point of view of American strategic thinking. After the disappearance of the Cold War the USA has become the sole surviving superpower and no longer faces a serious threat from one single formidable enemy, such as from the USSR earlier. The USA would understandably like to remain the single unchallenged global superpower, and its strategic thinking is guided by this objective. Given that objective, it may well feel the need to take into account the possibility of future trouble from diverse regional powers in different parts of the world, which might interfere with the perceived American interests in those areas. The objective could then be to see that such threats do not emerge in any part of the world. That is perhaps a rather crude summary of American thinking, but not, one hopes, too inaccurate.

It is in this context that the notion of 'regional security' assumes importance. The occurrence of conflicts over scarce natural resources or over environmental concerns in different parts of the world could conceivably seem an adverse development from the point of view of American strategic thinking in relation to the area in question, and if so, it may appear to warrant intervention in some form. This could be one of the considerations leading to an interest in such conflicts in different parts of the world on the part of the American military establishment, and the subsuming of such conflicts under the rubric of 'regional security'.

The author recalls a conference on the eastern Himalayan rivers held a few years ago, at which a distinguished American professor, referring to the Indo–Bangladesh dispute over Ganga waters, made an observation roughly on the following lines: 'If you people think that you can quarrel here without the rest of the world taking notice, you are mistaken; we will intervene and bash your heads together.' It was a rather crude exposition of *pax Americana* that no one took seriously, but it was not without its significance. It is this kind of interventionism that might be facilitated and encouraged by the postulation of a security angle to conflicts over scarce natural resources or environmental concerns, and the further categorization of this as a question of 'regional security'. It may not be the intention of scholars working on natural resource scarcities to facilitate or encourage such developments, but the terminology used might tend to lend academic support to the strategic thinking and planning of political administrations and military establishments.

Summing Up

Summing up, the following propositions emerge from the foregoing analysis:

- There are, and will be, scarcities of natural resources and environmental impacts of developmental activities, and these may lead to conflicts both within countries and between countries; some of these conflicts may be acute and violent.
- However, these conflicts are likely to be manageable and amenable to resolution; they are unlikely to lead to wars. The thesis of 'water wars' seems implausible. What is more probable is that there will be treaties, agreements, joint institutional arrangements and so on.
- Environmental concerns in relation to these matters are just as likely to lead to positive cooperation for mitigation and improvement, as to conflict.
- The grave environmental degradation and the consequent crisis facing humanity is global, not regional; the essential conflict here is not between a few developing countries in some parts of the world, but between the affluent 'advanced' countries and the rest of the world. This is obscured by the description of relatively minor conflicts relating to scarce natural resources or local environmental concerns as conflicts caused by 'environmental change'.
- The subsuming of eminently soluble local conflicts relating to scarce natural resources or environmental issues under the rubric of 'security' brings in a wrong orientation and wrong actors into the scene, and makes conflict-resolution more difficult and harmony less likely to achieve.
- The further categorization of this as a question of 'regional security' is fraught with the danger of bringing such conflicts within the ambit of post-Cold War American concepts of 'regional trouble spots' and their possible bearing on the perceived American interests in the regions concerned. Whatever the justification for that kind of strategic thinking on the part of certain establishments and agencies, the academic community should be wary of being co-opted into providing a theoretical underpinning for it.

Postscript

This chapter sounds a caution against the importation of the language of security into discussions of the scarcity of water and other natural resources, and deprecates the theory of water wars. That language and that theory, the chapter argues, are aimed at (or will have the effect of) facilitating intervention in various regions of the world by the sole super-power in the pursuit of its own perceived interests of diverse kinds. This does not imply a disagreement with Vandana Shiva who has written a book entitled *Water Wars*. By 'wars' she means conflicts of different kinds arising from the depletion, destruction and pollution/contamination of scarce natural resources, particularly water, and the natural environment. 'Water wars' in her sense can arise between the rich and the poor, between the rural and urban areas, between the urban middle class and tribal communities, between the state and civil society, between the community and the corporate sector, between local people and multi-national giants, between 'development' and nature, and of course be-tween political units within a country or between countries. She is particularly concerned about corporate 'wars' against the people, i.e., the taking away of both individual and community rights to land and water by the state and/or by the domestic corporate private sector, and the subordination of all these rights, as well as those of the state itself, to those of foreign or multi-national corporations under the regime of the WTO. The dangers that she cautions us about are real and need to be guarded against.

18

Water and Security in South Asia

A Cautionary Preamble

This chapter (based on a paper initially written for a conference on the environment/security linkages) is an attempt to examine what the notion of 'water security' can mean in the context of India's relationship with its neighbours.

Let us first get one obvious point out of the way. It is of course possible to use a natural resource or an environmental factor as a weapon of war or an instrument of political pressure. To give some fanciful hypotheses, a country may try to poison the water source of another country, or bomb a dam in an enemy country and create devastation and so on. These are straightforward security concerns in the conventional sense; the fact that natural resources or environmental factors (and not the usual weapons) are used as the means of inflicting damage is merely incidental. Again, where a river runs along, or marks, the boundary between two countries, there may be security concerns relating to that border similar to those relating to a land border. These are not the kinds of things that we are concerned with here; what we have in mind is the possibility of conflicts arising from scarcities of natural resources, their pollution or contamination through use, environmental degradation and so on. It is in relation to such conflicts, existing or potential, that the language of security is often invoked.

This has already been questioned in Chapter 17. To recapitulate the argument briefly, there is a fundamental confusion here between the different senses of the word 'security'. The mere fact that the same word 'security' is used in diverse contexts should not mislead us into thinking that we are talking about the same thing, and transferring concepts and approaches from one context into another. Further, 'security' in the military, intelligence or police sense is essentially a negative, adversarial concept. It consists of the perception and analysis of threats and the formulation of counter-measures, defensive or offensive. All this is far removed from the language of harmony or cooperation. When we are considering scarce natural resources or environmental concerns, we are in an altogether different universe of discourse. Here we have to think in terms of cooperation, harmony and collaboration; and in terms of natural systems, not national boundaries. Conflicts exist, of course, or may emerge, and they have to be resolved; but we would do well to use precisely that language, namely, 'conflicts' and their 'resolution'. When we begin to talk about these matters in the language of security, the entire tenor of the discussion changes into an embattled, defensive, distrustful, adversarial mode. A further danger is that the wrong actors enter the arena. The use of the terminology of security makes the 'security community' feel that this is *their* area, and this makes the processes of conflict-resolution more, and not less, difficult. In brief, wrong terminology provides a wrong framework, brings the wrong people in and pushes us in a wrong direction.

Some may say: 'We are not using the word "security" in a narrow, adversarial sense. We are in fact trying to enlarge the concept to cover a wider range of concerns, and to give it a *positive* slant.' But why torture the concept and force it to take on meanings that are not natural to it? However hard we try to sanitize the word and replace its negative, adversarial connotation by a positive, consensual one, a residue of its original sense and undertone will remain, and will be carried into (and influence) the newer applications. Our efforts to widen the mental horizons of the traditional security community and to sensitize them to newer concerns may or may not influence *their* thinking, but our borrowing of their terminology will surely influence *ours*.

The Area of Concern

Having entered those reservations, one must recognize that one cannot hope to change current usage. Let us then proceed to deal with the issue as posed. In the context of water, we are concerned mainly with the Indian subcontinent. Island countries such as Mauritius and Sri Lanka are not linked to the water systems of the mainland. For the present, at any rate, we can leave out China too, for the reason that it is physically separated from the Indian landmass by mountains and plateaux.[12] We are therefore talking about Pakistan, India, Nepal, Bhutan and Bangladesh, or to put it in terms of river systems, the Indus system in the west and the Ganga–Brahmaputra–Meghna (Barak) system or systems[13] in the east. (The abbreviation 'GBM' has come into use in recent years.)

Both in the west and in the east, the river systems were cut across by the lines of the Partition of 1947, transforming them into trans-boundary systems requiring inter-country understandings. It will be noticed that the reference is essentially to the northern half of the subcontinent. The southern peninsular part is unrelated to the Himalayan river systems, and while there have been and continue to be conflicts over the southern rivers (the Godavari, the Krishna, the Cauvery and others), there are no international dimensions to such conflicts.[14]

Against that background, what can 'security' mean in relation to water in the area that we are talking about? We can best answer that question not by trying to define the term, but by trying to capture the kinds of concerns that lie behind that usage. Approaching the subject in this manner, we can identify three broad categories of 'security' issues:

- water needs for diverse purposes (availability, adequacy, reliability, dependence, vulnerability, etc.,);
- the danger posed by floods: need for mitigation, management, damage-minimization;
- water quality problems: prevention or control of pollution and contamination.

---

12 The Brahmaputra (the Tsangpo in Tibet) flows largely through uninhabited or sparsely inhabited areas until it crosses southwards into India. There were reports some years ago that China was planning to divert the Brahmaputra northwards with the aid of nuclear explosions. One does not know whether there are in fact any such plans. If there are, that would be a matter for serious concern on the part of India and Bangladesh. However, for the time being there are no water-related issues between China and India.
13 There is a difference of views between India and Bangladesh on whether these three rivers constitute one basin or system or three different systems.
14 International dimensions could arise if any attempts were made to divert waters from the Himalayan rivers to the peninsular ones. There has been some vague, unrealistic talk of such possibilities, but they seem very unlikely to become serious propositions in the foreseeable future.

If we identify the problems in this manner, what needs to be done is clear enough:

- understandings on water-sharing on common river systems;[15]
- cooperation in the establishment and operation of effective and timely information and warning systems in regard to flood flows, and in disaster-preparedness and damage-mitigation; a coordination of coping strategies; a sharing of experiences in this regard;
- the institution of common standards on water quality, and understandings in regard to the maintenance of water quality across borders.

It will be noticed that 'energy security' is not referred to in the above enumeration. It seems to the author that with hydroelectric power—which is what we are concerned with in the context of water—the question is not so much one of security or insecurity, as one of a 'potential' to be utilized.

Floods

Floods and the proper response to them is a vast, complex and controversial subject that will need a separate paper. For our present purposes what we need to note is that floods have been part of the experiences of the countries in the Indian subcontinent from time immemorial and will continue to remain so in the future. The extent and duration of flooding and the damage that this causes are influenced both by *natural phenomena* (prolonged and heavy rainfall, intense precipitation during a limited period, sudden cloudbursts, the breaking of temporary barriers created by landslides or earthquakes, the occasional simultaneous occurrence of high floods in different river systems, e.g., the Ganga and the Brahmaputra, which normally have their peaks at different times, and the further

---

15 In referring to rivers that flow through more than one country, one has to use words such as 'common' or 'shared' but even these can give rise to controversy: at one stage there was a strong reaction in some quarters in Nepal to the use of the term 'common rivers' in a Memorandum of Understanding between the Prime Ministers of Nepal and India!

synchronization of this with tidal bore conditions in the estuary, as happens occasionally in Bangladesh, and so on) and by *human activity* (deforestation of hillsides, structures impeding river-flows, human settlement and economic activity in the flood plains of rivers, drainage congestion, and so on). Some of the efforts at flood control (embankments, dams) have themselves caused further problems on occasion. There is a growing realization that 'flood control' is not a feasible proposition; that embankments have often proved a remedy worse than the disease; that large dams are not often planned with flood moderation as a primary aim, and even where they are, the competing claims of irrigation and power-generation often override the flood moderation function; that dams, while they might moderate flood flows to some extent under normal conditions, might aggravate the position if water has to be suddenly released in the interest of the safety of structures; and that what we must learn to do is not so much to 'control' floods as to cope with them when they occur and minimize damage. The most important element here is timely knowledge and preparedness: this is the meaning of 'security' in this context. Countries (governments, people) need to know how soon a flood is likely to arrive, and what its magnitude is likely to be. They can then take appropriate measures for the prevention or minimization of hardship, loss and damage, and for relief where necessary. A vast, well-equipped, technologically advanced network of stations for observing and analysing precipitation and flows and drawing conclusions, and for the instant (real time) communication of such information and predictions to downstream areas and countries, is needed. This is an important area for inter-country cooperation. Not only knowledge, but experience and lessons learnt, need to be shared, and coping strategies exchanged and coordinated. Modest beginnings have been made in these directions but much remains to be done. (See Adhikary et al., 2000, Chapter 6, 'Cooperation in Flood Disaster Management'; and *Seminar*, July 1999).

Water Quality

This is another important area of concern. Here again, it is common knowledge that there has been rapid and grave deterioration of water quality in both surface-water and groundwater aquifers in all the countries of the subcontinent. All countries have pollution control laws and institutions, but these have not been able to prevent the growing pollution and contamination of water sources and systems, which in effect makes much of the 'available' water resources unusable. This is in fact as great a threat (if not greater) to security as the 'scarcity' about which alarm bells have been ringing. Much of the action here—prescription and continuous review of standards, their enforcement, not forgetting the cumulative impact of individual clearances and permits, making the polluter pay, adopting and moving towards clear, time-bound goals in regard to desired water quality—lies within each country, but pollution and contamination are not respecters of political boundaries. The countries of the region have to agree on common standards and on trans-boundary water quality protocols. Conflicts have arisen in the past over water-sharing, but water quality may well become the focus of even sharper conflicts in the future unless clear inter-country understandings are reached and appropriate institutional mechanisms are provided for ensuring compliance with such understandings. (See Adhikary *et al.*, 2000, Chapter 3, 'Water Quality in the GBM Region'.)

Having very briefly dealt with those two aspects of security, namely, flood management and water quality, let us now turn to what is generally perceived as the principal issue, namely, water-sharing.

Water-Sharing

India–Pakistan

There is already an understanding between India and Pakistan in so far as the Indus system is concerned: the Indus Treaty of 1960. (Please see Chapter 19, section V for a brief discussion on this.) The Treaty has been working reasonably well despite the difficult political relationship between the two countries; it was not abrogated even during periods of war. There are of course problems in the Indus basin, such as waterlogging, salinization, drainage congestion and so on. These cannot be attributed to the Treaty; they are the result of faulty water management and irrigation practices in both countries.

It follows from the above that 'water security' is not an issue between India and Pakistan, so long as the Indus Treaty continues to work. The growth of population, the pace of urbanization and the progress of what we call 'development' may indeed impose increasing pressures on the finite availability of water, but this will call for appropriate remedial measures within either country. The answer lies in better resource management, and not in talking about 'security'. The operation of the Indus Treaty has been relatively smooth precisely because it has somehow been insulated (partially if not wholly) from the difficult political and military relationship between the two countries, i.e., from the 'security' angle. One must hope that this will continue. (Please see the note at the end of the section on the Indus Treaty in Chapter 19.)

India–Nepal/Bhutan

There are really no serious water-security issues between India and Nepal or between India and Bhutan. What has been under discussion between India and these countries is the building of projects for hydroelectric power, irrigation and flood-moderation purposes. The point to be noted in the present context is that there are no questions of security or in-security here. (There are indeed some pending issues regarding water-sharing between India and Nepal in relation to certain rivers, particularly in the context of the Mahakali Treaty of February 1996; please see Chapter 19, section V.)

India–Bangladesh

It is only in the India–Bangladesh context that the expression 'water security' acquires some significance. Bangladesh is a water-abundant country, but it experiences seasonal shortages (and of course seasonal excesses, i.e., floods). As the lowest riparian in the Ganga–Brahmaputra–Meghna system(s), Bangladesh could well feel a certain sense of insecurity in relation to water, particularly considering the size of the immediate upper riparian, India, in terms of land and population. Bangladesh is very conscious of the fact that there are 54 rivers (including rivulets and streams) crossing the Indo–Bangladesh border, and that 94 per cent of its waters originate beyond its boundary. Clearly, it needs some reassurances regarding water. Fortunately, the most contentious (and at one time seemingly intractable) issue between India and Bangladesh, namely, the Ganga waters issue, which was a major component in the complex and difficult relationship between the two countries, stands resolved by the Ganges Water Sharing Treaty of December 1996 (see Chapter 19 section V). Agreements or Treaties are needed on some of the other rivers (Teesta, Muhuri, Manu, Gumti, Khowai, Brahmaputra, Dharla and Dudh Kumar) that have been identified as important. Water-sharing talks are proceeding on the Teesta river on which both countries have built barrages. These need to be brought to a quick conclusion, and followed by talks relating to the other rivers.

Inter-Country Cooperation

In the light of the foregoing, the question of 'water security' in the Indian subcontinent turns out to be a fairly limited one, requiring a few more agreements between India and Bangladesh. However, this may be dismissed as a 'reductionist' or 'minimalist' view of security. Some (in India as well as in Bangladesh and Nepal) will argue that if we take a wider view of possibilities of cooperation among the countries of the GBM system(s), large vistas will open up before us. That may be true, but we are no longer in the universe of discourse of 'security'. There is no particular reason why possibilities of cooperation, harmony and integrated development should be discussed under the rubric of 'security'.

V
Relations with Neighbours

19

Conflict-Resolution: Three River Treaties

Introductory

This chapter is about conflict-resolution in South Asia in the area of water resources. It is structured around three Treaties: The Indus Treaty of September 1960 (India–Pakistan), the Mahakali Treaty of February 1996 (India–Nepal) and the Ganga Treaty of December 1996 (India–Bangladesh). It covers very briefly the background to, and nature of, the dispute in each case, the approach to a resolution, the major features of the Treaty, the manner in which it has been operating, the difficulties encountered, how these can be resolved, etc. It then proceeds to set forth some explanations and reflections that arise from these cases.

Two clarifications may be in order at the outset. First, the primary concern of this chapter is with the governing perceptions and other operative factors, as well as the issues that came up, rather than with the details of the Treaties, though details are gone into as needed. Secondly, an important objective of the paper is the wider dissemination of knowledge beyond the limited circles to which knowledge about these matters is generally confined, without drawing upon classified sources.

The Indus Treaty 1960

In 1947 the line of Partition of the Indian subcontinent cut across the Indus system. There had been considerable irrigation development in the undivided Punjab based on the waters of the Indus system. This was disrupted by Partition and the large-scale movement of people. An understanding on water-sharing between the two new countries formed by Partition was clearly necessary. It was also necessary to facilitate the development of irrigation systems in the western part of Punjab that went to Pakistan. After prolonged talks between the two Governments, the constructive approach of Nehru and Ayub Khan, assisted by the good offices of the World Bank, led to the signing of the Indus Treaty in September 1960.

The water-sharing under the Treaty was quite simple: the three western rivers (the Jhelum, the Chenab and the Indus itself) were allocated to Pakistan, and the three eastern rivers (the Ravi, the Beas and the Sutlej) were allocated to India. Certain restrictions were placed on India as the upper riparian. On the rivers allocated to Pakistan, India was not allowed to build storages. Restrictions were also imposed on the extension of irrigation development in India. (On Pakistan, the lower riparian, there were some relatively less significant restrictions). There were also provisions regarding the exchange of data on project operation, extent of irrigated agriculture and so on. The Treaty further mandated certain institutional arrangements: there was to be a permanent Indus Commission consisting of a Commissioner each for India and for Pakistan, and there were to be periodical meetings and exchanges of visits between the two sides. Provisions were included for conflict-resolution: differences, if any arose, were to be resolved within the Commission; if agreement could not be reached at the Commission level, the dispute was to be referred to the two Governments; if they too failed to reach agreement, the Treaty provided an arbitration mechanism. The settlement also included the provision of international financial assistance to Pakistan for the development of irrigation works for utilizing the waters allocated to it, and India too paid a sum of £ 62.06 million in accordance with Article V of the Treaty.

It could be argued that the division of the river system into two segments, one for Pakistan and one for India, was not the best solution, and that there should have been a sharing on all the rivers, or a joint integrated planning and management of the totality of the system by the two countries. However, such possibilities were probably ruled out by the state of relations between the two countries, and the Treaty doubtless represented the best arrangement that was negotiable at the time.

The Treaty has acquired a reputation internationally as a successful instance of conflict-resolution, and is often hailed as such in the literature on the subject. It has been widely noted that it has been working reasonably well despite the difficult political relationship between India and Pakistan, and that it was not abrogated even during periods of war between the two countries. The Indus Commission has been meeting regularly in either country, and the working relationship between the engineers at the Commission level is very cordial. Differences do arise from time to time, but these usually get resolved within the framework of the Treaty. Minor differences are settled within the Commission, and major disputes go to the two Governments. So far, it has not been found necessary to invoke the provisions for arbitration by a third party. An important dispute that arose during the 1970s was regarding the Salal Hydro-Electric Project in Jammu and Kashmir. This was referred by the Commissioners to the two Governments, and after lengthy and difficult negotiations the issue was eventually resolved.

At present there is an unresolved dispute regarding what is known as the Tulbul Navigation Project (or the Wular Barrage Project) on the Jhelum river. Pakistan objects to this project on the ground that it involves the creation of a storage on a river allocated to Pakistan and that it is therefore a violation of the Treaty. India argues that no creation of storage is involved; that the proposed barrage will merely head up the waters temporarily, retarding the rapid depletion of flood waters, with a view to extending the period during which navigation is possible; and that the regulation involved will also benefit Pakistan. The intergovern-mental talks on the subject have not so far been successful. There is also another long-pending dispute regarding a run-of-the-river project (Baglihar). Here again there is a divergence of perceptions between the two countries as to the conformity of the project to the Treaty. These disputes too, like the Salal dispute, may have been settled between the two Governments in due course if the circumstances had been normal. Unfortunately, the relationship between the two countries has been very difficult in recent years, and so these disputes have remained unresolved.

(Note: The above was initially written in 1999–2000. In the two years since then the political relationship between the two countries has greatly deteriorated. That strongly negative relationship, which is per-haps at its worst ever now, has had its impact even on the functioning of the Indus Treaty. There was some speculation in the media about the possible abrogation of the Treaty, but it is not clear whether such a course was seriously contemplated. The Indus Commission did meet at Delhi on 29 May 2002 as scheduled. If there was in fact a crisis it seems to have blown over. For the time being, at any rate, the Treaty seems to have managed to survive the strain this time also. However, outstanding disputes are of course unlikely to get resolved in the near future. For a more detailed discussion of the subject, refer to the author's article 'Was the Indus Waters Treaty in Trouble?' in the issue of *Economic and Political Weekly* dated 22 June 2002.)

'The Eastern Himalayan Rivers'

Turning to the east, the term 'eastern Himalayan rivers' that has gained considerable currency in recent years is a slightly imprecise expression, but it is generally known that it refers to the rivers Ganga, Brahmaputra and Barak or Meghna. Nepal has several rivers, all of which are a part of the Ganga system; that country has nothing to do with the Brahmaputra or the Barak. Bhutan, on the other hand, is entirely within the Brahmaputra system. Both India and Bangladesh are concerned with all three rivers. In India, the Ganga, the Brahmaputra and the Barak are far apart, whereas in Bangladesh they all join and flow together into the sea; and yet Bangladesh argues that they are three separate basins and India holds the view that they constitute one basin because of the common terminus. Be that as it may, together these systems (or this system) represent(s) vast water resources. In relation to these water resources the countries concerned (leaving China aside for the present) can be classified as follows: Nepal/Bhutan: upper riparians, mountainous countries; India: middle riparian (lower riparian to Nepal/Bhutan, upper to Bangladesh), largely plain country, though there are some hilly areas; Bangladesh: lowest riparian, deltaic country.

To Nepal and Bhutan, given their relatively limited cultivable land areas, their water endowment largely represents a source of hydroelectric power, whether for use within the two countries or for earning revenues through export[16]; to Nepal, the rivers also hold promise of an escape from its landlocked condition through a navigable outlet to the sea. To India, the water resources represent possibilities of hydroelectric power, navigation and irrigation, but also the danger of floods. To Bangladesh, criss-crossed as it is by a vast network of rivers and streams, they represent

---

16 While the mainstream view in Nepal is that water is to Nepal what oil has been to the Middle East, namely, a source of revenue—through the export of energy from large hydroelectric projects—there is another school of thought in that country that argues for people-centred development through local schemes and projects on a small and medium scale which are sustainable and in harmony with nature. The voice of the latter, however, is not the dominant one.

a formidable source of recurrent and often devastating floods, but they also act as conveyors of commerce, sustain fisheries, offer possibilities of irrigation and play a role in the control of salinity.

Let us now look a bit more closely at some of the conflicts that have actually arisen and been resolved.

India–Nepal Issues

Past History

In so far as India–Nepal relations are concerned, water-sharing has not been a major issue. There has been no conflict over water. (What appears to be a question of water-sharing has now come up in the case of the Mahakali; we shall return to this). What is involved in the Indo–Nepal (as also in the Indo–Bhutan) context is essentially a question of cooperation in deriving benefits from the water resources by way of hydropower, irrigation, flood management and perhaps navigation. However, the very attempts at cooperation can give rise to conflicts, and have actually done so.

Two early projects, far from promoting good relations between the two countries, caused serious strains in that relationship. The Kosi/Gandak agreements were not regarded at the time as exercises in 'regional cooperation'; that term came into use much later. These were essentially projects conceived by India to meet its requirements or solve its problems, with some benefits to Nepal included. That was the way the projects were designed with Nepal's agreement, but they were subsequently criticised in Nepal for conferring substantially more benefits on India than on Nepal, though this was inevitable given the relative magnitude of cultivable areas in the two countries. The projects also suffered from poor design, inefficient implementation and bad maintenance; even what was promised was not delivered, either in Nepal or in India. The Kosi/Gandak agreements, initially signed in 1954/1959, were amended in 1966/1964 to take care of Nepalese concerns, but the sense of grievance was not wholly removed.

The bitterness generated by these experiences coloured all subsequent dealings between India and Nepal. Suspicion and mistrust grew and became a massive impediment to good relations between the two countries. It became *de rigueur* for all Nepalese commentators to blame India for playing 'Big Brother' in relation to Nepal. That India 'cheated' Nepal with regard to the Kosi and Gandak rivers, and was bound to do so on other matters, became established Nepalese belief. Occasional blunders and stupidities by India lent a degree of credibility to such accusations. No Nepalese politician was willing to take the risk of signing or supporting any kind of treaty or agreement with India; and the ruling political party at any given time was liable to be severely criticised by other parties in the event of its reaching an understanding with India on any matter. Against that background, none of the projects—Karnali, Pancheshwar, Saptakosi—which had been under discussion between the two countries could make any real headway. The mistrust and suspicion led to the inclusion of a provision in Nepal's Constitution making Parliamentary ratification by a two-thirds majority necessary for any treaty or agreement relating to natural resources and likely to 'affect the country in a pervasively grave manner or on a long-term basis'; in the case of treaties of 'an ordinary nature' ratification by a simple majority was laid down[17].

A further chapter was added to the old history of misunderstanding by the Tanakpur episode. In itself it was a minor matter of a small piece of land—2.9 hectares (ha) to be exact—being used by India for building the eastern afflux bund for the protection of Nepalese territory from possible backwater effects from the Tanakpur barrage (which itself was wholly in Indian territory and did not involve the consumptive use of water). The initial unwisdom on the part of India in not keeping Nepal informed in advance about this project was subsequently sought to be remedied by reaffirming Nepalese sovereignty over the aforementioned territory (as well as the pondage area of 9 ha), and agreeing to provide some free electricity and water to Nepal, the quantum of which was revised upwards twice in inter-governmental understandings. However, Tanakpur came to loom large in the Nepalese consciousness and was used as an issue for bringing down the Koirala Government. It was in fact more of an issue in domestic politics than an Indo–Nepal controversy, though it had the potential of souring Indo–Nepal relations. The Supreme Court of Nepal, in response to a petition, ruled that the Memorandum of Understanding between the two Prime Ministers on Tanakpur was indeed in the nature of a Treaty or Agreement, but left it to the Executive Government and Parliament to decide whether

---

17 For a more detailed account see Chapter 13 of Verghese and Iyer (1993).

ratification by a simple or two-thirds majority was needed. This was not put to the test; instead the Tanakpur controversy was subsumed in negotiations over a larger issue.

The Mahakali Treaty

Political relations between India and Nepal, which had become strained during the Rajiv Gandhi period, had begun to improve significantly with the advent of Parliamentary democracy in Nepal and changes of Governments in India. Despite the Tanakpur *contretemps*, this trend continued. A new chapter in Indo–Nepal relations seemed to open with the Mahakali Treaty of February 1996. The signing of the Treaty was preceded not merely by negotiations between the two Governments but also by extensive informal consultations covering all parties in Nepal so as to facilitate the process of parliamentary ratification. After much difficulty and suspense, parliamentary ratification by a two-thirds majority also came through. Now the Treaty is in force and in the process of implementation, but that process has been stalled by some differences. However, before going into those problems, it is necessary to take a quick look at the contents of the Treaty.

The Treaty was 'Concerning the Integrated Development of the Mahakali River'. The preamble described the Mahakali as 'a boundary river on major stretches between the two countries', and the Treaty itself as a 'treaty on the basis of equal partnership'. The Treaty covered the Sarada Barrage[18], the Tanakpur Barrage and the proposed Pancheswar project, and replaced earlier understandings on these matters. From the Sarada Barrage, the Treaty gave Nepal 1,000 cubic feet per second (cusec) of water in the wet season and 150 cusec in the dry season, and provided for this water to be supplied from Tanakpur in the event of the Sarada Barrage turning non-functional; it also included the interesting provision that not less than 350 cusec should flow downstream of the barrage to maintain and preserve the ecosystem of the river. On Tanakpur, the Treaty reaffirmed Nepalese sovereignty over the land (2.9 ha) needed for building the eastern afflux bund, as well as the 9 ha of pondage area. 'In lieu of the eastern afflux bund' (presumably this means 'in

---

18 The Mahakali is the name of the river in Nepal; after it crosses into India it is known as the Sarada.

consideration of ...'), the Treaty gave Nepal the right to 1,000 cusec of water in the wet season and 300 cusec in the dry season; and 70 million kilowatt-hours (kWh) of electricity (as against the earlier agreed figure of 20 million kWh). As and when the Pancheswar Project came into being and augmented the availability of water at Tanakpur, Nepal would be provided with additional water and additional energy, with Nepal bearing a part of the cost of generation of incremental energy. There was a provision for the supply of 350 cusec for the irrigation of the Dodhara–Chandani area. On the Pancheswar Project, which was to be located on the Indo–Nepal boundary and was to be a joint project, some general principles applicable to border rivers (of which an important one was 'equal entitlement in the utilization of the waters of the Mahakali River, without prejudice to their respective existing consumptive uses of the waters') were laid down, which were further elaborated in a side letter exchanged by the two Prime Ministers. The detailed project report (DPR) was to be jointly prepared in six months; the energy, irrigation and flood control benefits to the two countries were to be assessed, and the capital cost shared accordingly; the power benefit was to be assessed on the basis of savings in costs as compared with the relevant alternatives available; and so on. There was to be a binational Mahakali River Commission, guided by the principles of equality, mutual benefit and no harm to either party. There would also be a specific joint entity to develop, execute and operate the Pancheswar Project. There were other provisions relating to the life of the Treaty (75 years), review after 10 years, arbitration in the event of disputes, etc.

As already mentioned, the Treaty is formally in operation now, but the progress in its implementation has been tardy. The DPR which was to have been prepared in six months got stalled partly because of certain technical differences (for instance, regarding the location of the re-regulating structure downstream of the dam), and partly because of some more serious differences of a political nature. There is in fact an *impasse*, though there are some signs of movement at long last. Let us consider the problems briefly.

(i) The Parliamentary 'Strictures': First, the Parliamentary ratification of the Treaty was accompanied by a set of resolutions (*sankalp prastav*) that are referred to in Nepal as 'strictures'. If the Parliament of Nepal had been deeply troubled by certain questions, it could have refused to ratify the Treaty. To say that it did ratify the Treaty but at the same time passed a series of 'strictures' seems untenable. In any case, 'strictures' by the Nepalese Parliament can apply only to the Nepalese Government, and not to the Government of India. The Government of Nepal must of course take note of their Parliament's concerns, and if necessary, go back to the Government of India for a fresh round of negotiations; but in that event the Treaty must be treated as dormant (if not as non-existent) until the renegotiation is completed and a fresh document is agreed upon. And of course the negotiations may fail, or may yield results that the Nepalese Parliament may not approve of. It is possible to argue that there can only be 'ratification' or 'non-ratification' of a Treaty, and not a *conditional* ratification; that a conditional ratification is the same as non-ratification, and that the Mahakali Treaty does not stand ratified and therefore does not exist. However, no one argues on these lines; the general view in Nepal is that the Treaty has been ratified. At the same time, the 'strictures' tended—at least for some time—to cast a shadow on the Treaty and come in the way of a serious effort to deal with the differences that had emerged.

(ii) The Kalapani issue: Nepal questions the Indian military presence in an area called 'Kalapani' (near the source of the Mahakali River) on the ground that it is Nepalese territory. This is a territorial dispute. Either the area in question is part of Indian territory or it is not. If it is, the Indian military presence there is a matter of no consequence to Nepal; if it is Nepalese territory, India has no business to be there. This is a matter to be resolved with reference to old records, documents, maps, survey reports, etc. The dispute needs to be settled quickly in a spirit of goodwill, and not be allowed to fester. Nothing is gained by arousing emotion over this issue, and in any case, this has nothing to do with the implementation of the Mahakali Treaty. (It appears that this question is now being discussed separately and not in the context of the Mahakali Treaty.)

(iii) 'Boundary River': The Nepalese view, drawing support from the Parliamentary resolution, is that the qualification 'on major stretches' should be ignored and the Mahakali treated simply as a boundary river. No one can write that kind of a gloss on the Treaty. The words of a Treaty, which are the result of hard negotiation, are sacrosanct. If the Treaty says 'boundary river on major stretches' then that is what it is.

(iv) 'Equal sharing': From the fact that the Mahakali is a boundary river the Nepalese draw the inference that it belongs equally to the two countries, and therefore that half of the waters of the river belong to Nepal. These doctrines ('boundary river', ownership of half the waters, etc.,) seem to be Nepalese innovations not easily derivable from any international law or principles. The Indian view (if one has understood it correctly) is that the river can be used by the two countries but that it does not 'belong' to either; that in particular, any doctrine of ownership of flowing water and the implied right of the upper riparian to 'sell' the water so owned to the lower riparian (who would in any case receive that water naturally by gravity flow) seems non-maintainable in international law; that 'equal sharing' really applies to the incremental benefits to be created by the Pancheswar Project, and that the relative benefits gained by the two countries would determine their respective shares of the capital costs of the project. There is a clear divergence of views here. In so far as this is the result of inadequate negotiation or poor wording, both sides must share the blame for leaving this nebulousness in the Treaty. Nothing will be gained by taking a dogmatic position on this issue; this is a matter for discussion between the two countries with a view to arriving at an agreed position.

(It appears that the Government of Nepal, while explaining the merits of the Treaty to Parliament and the public, had made the questionable claim that the Indians had been persuaded to deviate from the Helsinki principle of 'equitable apportionment for beneficial use' and accept that of 'equal sharing'. If so, one can only say that the Government of Nepal has made its own task somewhat more difficult. No one in India thinks that any such deviation has been agreed to; how would anyone have had the authority to make any such deviation?)

(v) The protection of existing consumptive uses: Under the Treaty, the sharing of the capital costs of the Pancheswar Project would be in proportion of the relative incremental benefits, and the incremental benefits have to be reckoned after protecting existing consumptive uses of the waters of the Mahakali. India has claimed that there is such an existing consumptive use at the lower Sarada, but Nepal questions this on certain grounds: that this use had not been mentioned earlier; that the Treaty covers only the Mahakali River as a border river, and not the river after it crosses the border and becomes an Indian river; and that the farmers at the lower Sarada are really drawing upon the Karnali[19] and make only occasional use of the Sarada waters. This issue too can be easily resolved if answers are found to the following questions: (a) Is there an existing consumptive use of Mahakali waters in the lower Sarada area? If so, what is the quantum? How old is the use? Is it regular or occasional? Is it a fact that the farmers depend essentially on the Karnali river and draw upon the Sarada only infrequently, when for certain reasons they are unable to use Karnali waters, and if so, how important is that occasional use? (b) What would be the consequences of not recognizing this as 'existing use'? Is it merely a question of reckoning this against India's share of the benefits arising from the Pancheswar Project, and thus requiring India to pay more (perhaps a few hundred crores) towards the capital cost of the project? Or is there a danger of an actual denial of Mahakali waters to the farmers in question? (Incidentally, if this is in fact a case of prior use, would it not be entitled to consideration under the general international conventions and practices even if there were no Treaty?) (c) In the event of the farmers being denied Mahakali waters, will they have any alternative water source, or will they be subjected to distress? As a result of this examination it may possibly be found that there is no real problem, or that it is marginal, and that solutions are available; but it is necessary to study the matter first. (Note: It appears that the two countries are moving closer to an agreed solution on this issue.)

(vi) Power tariff: The side letter to the Treaty says that the power benefit is to be assessed on the basis of saving in costs as compared with the relevant alternatives available. Two questions would then arise: first, what in fact is the 'alternative', and secondly, should the tariff be the same as that of the alternative cost? In regard to the first question, there are many possibilities (other hydro-electric projects, thermal projects, gas-based projects, etc.,) and

---

19 The Karnali, after it crosses into India, is known as the Ghagra.

thermal generation need not be assumed to be the only alternative available. In regard to the second question, if in fact the generation cost at Pancheswar is lower, the gain would surely have to be shared between the two countries: if the 'alternative cost' is to be fully paid by India to Nepal, what is India's gain, and what has it 'avoided'? In any case, the price of power is not a question of abstract principles but one of negotiation: it will have to be attractive enough for Nepal to warrant the undertaking of a big project and affordable enough for India to warrant purchase from this source. Here again, the difference, if any, does not seem insurmountable; but so far as one knows, this question has not yet come up for serious discussion.

It will be a mistake to take a gloomy view of such differences and difficulties and to regard these as indications of the failure of the Treaty. What is important is that they should be quickly dealt with and settled. Delay and drift will render them more difficult and perhaps even intractable. Unfortunately, delay and drift were what seemed to be happening for some time, though there now appears to be some movement at last.

Incidentally, we must also note that despite the Parliamentary ratification, there does exist a body of opinion against the Treaty in Nepal. One does not know how widespread and influential this view is. An article strongly critical of the Treaty, as well as of the manner in which it was negotiated and then got ratified, appeared in the journal *Himal* (April 2001 issue), and a comment by the author was carried in the June 2001 issue. Some of the points made in the *Himal* article (by Gyawali and Dixit) need to be carefully considered. The author shares some of their concerns, and is not enthusiastic about large-dam projects in the Himalayan region. However, whatever our views about these matters, the fact is that there is a Treaty formally entered into by two countries; and it seems clear that both signatories want the Pancheswar Project implemented. That Treaty has got bogged down because of certain differences between the two Governments. Given that situation, we can treat the Treaty as a dead letter and rejoice in its presumed demise. Alternatively, we can take the view that any accord, however imperfect, is better than discord; that the failure of the Treaty will be fraught with serious consequences for the relationship between the two countries and that everything possible must be done to make the Treaty work to the advantage of both countries regardless of any reservations that one may have on some of the contents.

Indo–Bangladesh Issues

In the Indo–Bangladesh context, the major issue from the point of view of Bangladesh has been essentially one of water-sharing. There are 54 rivers and streams that cross the Indo–Bangladesh border, and questions of water-sharing can arise with regard to any one of them. The dispute regarding the sharing of Ganga waters has been settled. The next river on the agenda is the Teesta, on which a sharing agreement has become a matter of urgency as both India and Bangladesh have barrages on the river and are beginning to develop irrigation in their respective commands. On the sharing of the other rivers, some proposals of a general nature have been informally mooted, but there has been no serious discussion about them. Sooner or later, agreements may have to be reached on at least some, say six or seven, of these rivers. There have been some desultory efforts at cooperation over flood management, but much more needs to be done. In the present chapter we shall be mainly concerned with the Ganga waters dispute and its resolution.

The Ganga Waters Dispute

In the relationship between India and Bangladesh, the dispute over Ganga waters was for two decades an important component, perhaps the most important one; and though it now stands resolved by the Treaty of December 1996, it would be a mistake to regard it as having wholly disappeared. This chapter will not go into the history of the dispute in detail;[20] but it is necessary to take note of the nature and substance of the dispute.

A simplified version of the Bangladeshi view of this dispute would be as follows: that there was a 'unilateral diversion' of the waters of the Ganga by India at Farakka to the detriment of Bangladesh; that the resulting reduction in flows had severe adverse effects on Bangladesh;

---

20 See Verghese and Iyer (1993), Chapter 12 for a detailed account of the history of past Indo-Bangladesh negotiations on this issue; also Chapter 11 on the politics of riparian relations. See also Crow (1995).

and that this was a case of a larger and more powerful country disregarding the legitimate interests of a smaller and weaker neighbour, and callously inflicting grievous injury on it. That view of the dispute has been widely prevalent in Bangladesh, cutting across all kinds of divisions. A national sense of grievance grew and became a significant factor in electoral politics. In its extreme form the nationalistic position became a myth with India being cast in the role of a demon: whether Bangladesh was afflicted by drought or by floods, the responsibility was laid at India's door. 'Farakka' was blamed for all kinds of ills.

The perceptions on the Indian side were entirely different. At the governmental level, a fairly common view was that Bangladesh was extremely rigid and unreasonable on this issue; that it had greatly overpitched its water needs and was claiming a disproportionate share of the waters in relation to the relevant criteria (contribution, cultivable area, etc.,); that it tended to exaggerate the adverse effects of reduced flows; and that it had blown the dispute up into a big political issue in domestic politics, thereby making inter-governmental negotiations difficult. A further complication was that there was a feeling at the level of the State Governments in India that in its negotiations with Bangladesh the Central Government had failed to pay adequate heed to the interests of the States (some of which are very large entities) and tended to be generous to Bangladesh at their cost. This perception had a degree of importance in State-level politics, though the issue was not as important politically in India as it was in Bangladesh.

India tended unconsciously to regard the Ganga as essentially an Indian river, a river which in its view was short of water in terms of the cultivable area and the population served by it, and a river which had also to be reckoned as a part of the water resources available to the country, from which to meet the water needs of the arid areas in the western part of the basin and beyond, and perhaps even in the southern part of the country; in that national perspective there was little room for a serious consideration of the needs of Bangladesh. There was also inadequate appreciation of the ill effects suffered in Bangladesh because of the reduced flows in the Ganga. This myopia on the part of India was matched by a myopia on the part of Bangladesh too: a tendency to regard the undiminished continuance of historic flows as a birthright, a failure to recognise the needs of upstream populations, a refusal even to consider other possibilities of meeting the water needs of its own people for fear of compromising its claim over Ganga waters and a resentment and distrust of India as a big and powerful country. There were many, both in Bangladesh and in India, who held more nuanced and balanced views, but it was the oversimplified 'nationalistic' view that tended to carry greater weight in the bureaucratic and political processes in either country.

Unfortunately, the protracted inter-governmental negotiations on the subject were partly misdirected because of two red herrings that tended to distract attention from the central problem. The first was the Farakka barrage and the second the idea of 'augmentation'. The primary purpose of the Farakka barrage was the diversion of a part of the waters of the Ganga to the Bhagirathi–Hooghly arm for arresting the deterioration of Calcutta Port; and the secondary purpose was to protect Calcutta's drinking and industrial water supplies from the incursion of salinity. The construction of a diversion barrage across the Ganga was an idea that was first mooted by British engineers in the 19th century. Even if Partition had not taken place, the barrage would probably have been built. The concern which was felt in East Pakistan (later Bangladesh) was regarding the reduction of flows which would result from the construction and operation of the barrage. Since India too was entitled to use the waters of the Ganga (with or without the construction of a barrage), the issue was the quantum of flows that Bangladesh could insist on as its legitimate share. Thus the issue was not Farakka but water-sharing. It would eventually have come up between the two countries as upstream uses increased, even if the Farakka barrage had never been built. Farakka served to bring the issue to the fore much earlier than would otherwise have happened; at the same time the issue was obscured by wrong formulations. It became known as the Farakka issue. Whatever the wisdom and soundness of the project (and these have been questioned), the barrage once built was not going to be demolished. What was needed was an agreement on the quantum of water that would be allowed to go through the barrage in the lean season. Such an understanding did exist from time to time (1977–82; 1982–84; 1985–88). If there were a long-term agreement on a reasonable sharing of the waters during the lean period, there would be no 'Farakka problem' though the structure at Farakka would continue to exist. (Such an agreement exists now).

The second source of prolonged and fruitless controversy between the two countries was the idea of 'augmentation'. At a fairly early stage it was agreed at the level of the Prime Ministers of the two countries that the existing lean-season flow was inadequate for the combined requirements (as stated) of the two countries, and that it was necessary to 'augment' those flows by harnessing the water resources of the region available to the two countries. Unfortunately there was sharp disagreement on how exactly the augmentation was to be accomplished. The Indian proposal was for the augmentation of the water-short Ganga from the water-surplus Brahmaputra by means of a huge link canal from Jogighopa in Assam to Farakka, running right across Bangladesh. The Bangladesh proposal was for augmentation from within the Ganga system by storing its monsoon flows behind seven high dams in Nepal. Each side had serious reservations about the other's proposal. Endless discussions produced no agreement. The details of the proposals and the disagreements need not be gone into here.[21] We need merely note that the sharp divergence between India and Bangladesh on the modalities of augmentation led to an impasse on the water-sharing dispute. What the two countries needed to do was to put aside the augmentation controversy and try and arrive at a long-term agreement on the sharing of the lean-season flows of the Ganga. (This was what eventually happened in 1996).

Leaving those distractions aside, let us look at the crux of the problem. It was essentially one of water-sharing in the lean season (i.e., season of low flows), which had been identified in the 1970s as 1 January to 31 May. The lowest flow at Farakka, reached in the period from 21 April to 30 April, was determined by the India–Bangladesh Joint Rivers Commission (JRC) in the mid-1970s as 55,000 cusec (on a 75 per cent dependability basis, which means that the lowest flow is expected to be not lower than that number in 75 out of 100 years). If a sharing of the flow in the lowest 10 day period could be agreed upon, the sharing during the rest of the lean season would present no difficulty. Unfortunately, each country laid claim to the totality of the lowest flow for its needs. In 1977 the impasse was broken by an act of statesmanship on the part of Irrigation Minister Jagjivan Ram who, when the talks had almost collapsed, overruled his technical and other advisers and agreed to release as much as 34,500 cusec (which worked out to 63 per cent) out of the 55,000 cusec to Bangladesh, keeping only 20,500 cusec (37 per cent) as India's share. In subsequent Memoranda of Understanding (1982, 1985) this proportion was retained. In India, the general opinion was

---

21 See Verghese and Iyer (1993), Chapter 12.

that the allocation of 34,500 cusec out of 55,000 cusec was excessively generous to Bangladesh; that it was a purely temporary arrangement pending the institution of augmentation measures, and that it could not possibly be agreed to on a long-term basis. In Bangladesh the tendency was to regard 34,500 cusec as a conceded right and to treat it as a point of departure for further negotiations. This divergence was one of the factors that made agreement difficult.

The rigid positions taken by the two Governments made the negotiations difficult enough, and the difficulty was compounded by the political dimensions of the dispute. The deteriorating political relationship between the two countries rendered this problem an intractable one, particularly because it had by then become (or had been made into) a major issue in domestic politics in Bangladesh; and in turn the water dispute contributed to the deterioration of the relationship. After the MoU of 1985 lapsed in 1988 there was no agreement or understanding between the two countries for several years. There was also no serious effort to enter into discussions with a view to finding a lasting solution to the problem. (In the absence of any agreement, India gradually increased the diversion towards the Bhagirathi–Hooghly arm, leading to renewed complaints from Bangladesh).

The circumstances in mid-1996 seemed propitious to the conclusion of an agreement. The new Governments in position in Delhi and Dhaka were well disposed towards each other. There was a feeling on both sides that the *impasse* over the Ganga waters issue was unfortunate and that it was desirable to arrive at an early understanding on water-sharing, putting aside the augmentation question. High-level visits were exchanged and there was much talk of a 'window of opportunity'. However, there were difficulties that could not be made light of.

The Awami League in Bangladesh had in the past suffered under the disadvantage, in domestic politics, of being considered 'pro-India'. Their Government understandably found it necessary to demonstrate its nationalist credentials by seeking a better settlement from India than the previous governments were able to secure. Under those circumstances, the sharing pattern of the 1977 agreement became a floor on which it sought at least a token improvement.

This was bound to run into political difficulties on the Indian side, perhaps not as severe as those faced by the Bangladesh Government, but nevertheless real. Any increase in the releases ex-Farakka to Bangladesh would have necessarily meant a corresponding decrease in the availability of water for irrigation in West Bengal and for keeping Calcutta Port operational.[22] The West Bengal Government had to consider possible criticism by the Opposition parties that the interests of the State were not being safeguarded. Against that background, any proposal which envisaged an agreement with Bangladesh on the basis of the 1977 allocation, particularly a long-term one, was bound to present some difficulties to the State Government.

Many people in both countries were doubtful whether the gap between what Bangladesh might consider acceptable and what India might consider reasonable and feasible could be bridged.

The Ganga Treaty

In the event, the Treaty on the sharing of the waters of the Ganga entered into by India and Bangladesh on 12 December 1996 was a more significant document than most people had considered possible. Behind that success lay several factors: the patient efforts of the Indian Ministry of External Affairs from 1995 onwards to find a solution to the problem; the wisdom and courage of the Prime Minister of Bangladesh Sheikh Hasina in tackling an issue fraught with considerable political risks; the high priority that the new Indian Foreign Minister I.K. Gujral attached to the resolution of the Ganga waters dispute; the constructive and sagacious role played by the Chief Minister of West Bengal (Jyoti Basu) and the contributions made by his Finance Minister (Ashim Dasgupta) to the finding of answers to the difficult questions that came up in the negotiations; and last but not least, facilitatory efforts at non-official levels.[23] It must be added that quite understandably and necessarily, it

---

22 A practical and objective reassessment of the needs of Calcutta Port, in the light of a realistic appreciation of the kind of future it has, is not easy in the context of an emotional attachment to the Port and fears of large-scale retrenchment of the labour force.

23 During the period from 1990 onwards, when there was a virtual *impasse* between the two governments on this issue, the Centre for Policy Research (CPR), New Delhi, the Institute for Integrated Development Studies (IIDS), Kathmandu, and the Bangladesh Unnayan Parishad (BUP), Dhaka, were collaborators in a study on the optimal utilization of the eastern Himalayan rivers for the benefit of India, Nepal and Bangladesh, and among the subjects studied was the Ganga waters dispute. A paper by the author outlining an

was a political rather than a technocratic settlement. As an exercise in the reconciliation of differences the Treaty was a remarkable and ingenious document, as can be seen from the following brief *résumé* of its major features.

Duration: Bangladesh's demand was for a Treaty or a long-term Agreement. On the other hand, there were misgivings on the Indian side about making long-term commitments on the sharing of scarce resources in the context of growing demands. These two divergent concerns needed to be reconciled. The Treaty manages to do this. It is a 30-year Treaty (renewable further), with a provision for a review at the end of five years, or even at the end of two years, if either party wants it.

Water-sharing: The divergence over 'augmentation' which had led to an impasse in the past has been sidestepped: the Treaty is essentially regarding the sharing of lean-season flows, though there is an article which recognizes the need for cooperation in finding a solution to the long-term problem of augmentation. The sharing formula given in Annexure I to the Treaty is related to *actual* flows at various levels and not to '75 per cent dependable flows' as in past agreements; this simplifies matters. (The application of the sharing formula is demonstrated in Annexure II with reference to 'average' flows based on data from 1949 to 1988, but this is purely for illustrative purposes). The basic formula is that of equal (50:50) sharing of the lean-season flows by the two countries, which is as fair a principle as any. This applies to a range of flows, with two modifications at the upper and lower ends respectively. At the upper end, there is a slight acceleration of the increase in India's share to enable it to reach 40,000 cusec (the full diversion capacity of the Farakka feeder canal) at a flow level of 75,000 cusec instead of 80,000 cusec. Above 75,000 cusec, India's share is held at 40,000 cusec and the balance goes to Bangladesh. At the lower end, the basic 50:50 sharing is subject to the proviso that in the leanest part of the lean

---

approach to the resolution of the dispute and putting forward a set of proposals was extensively discussed at seminars and conferences, official and non-official, at Delhi and Dhaka in the years 1995 and 1996. There is reason to believe that these efforts at a non-official level (often referred to as 'Track II') did make a modest but useful contribution to the search for an agreement. There was in fact considerable interaction between 'Track I' and 'Track II' from the latter half of 1995 to December 1996.

season—from 11 March to 10 May—each side will be given a guaranteed 35,000 cusec, with the residue going to the other side, but in alternate 10-day periods (three 10-day periods in India's favour and three in Bangladesh's).[24]

Safeguards for Bangladesh: One of the demands of Bangladesh was that the Treaty should provide for a guaranteed minimum flow. (There had been a 'guarantee clause' in the 1977 agreement, and a 'burden-sharing' formula in the 1985 documents). The Treaty does not include a 'minimum guarantee'; however, it has several scattered provisions which together provide a measure of security to Bangladesh. First, there is the guarantee (already referred to) of 35,000 cusec to either side in alternate 10-day segments in the period from 11 March to 10 May. Secondly, when the flow falls below 50,000 cusec, the Treaty recognizes an emergency situation and provides for immediate consultations by the two Governments. Thirdly, as increasing upstream uses may over a period of time result in reduced flows in the river, the Treaty requires the Government of India to make every effort to protect the flows arriving at Farakka. Finally, in the context of the provision for a review at the end of five years (or even at the end of two years if either party desires it), the Treaty lays down that pending agreed adjustments as a result of such a review India shall release to Bangladesh not less than 90 per cent of its entitlement under the Treaty.

Conflict-resolution: The Treaty provides essentially for mutual consultations. There will be a joint monitoring of flows, which should eliminate or minimize the possibility of disagreements over numbers. Conflicts or disputes (if any) in the course of the operation of the Treaty will be resolved within the Joint Committee envisaged by the Treaty, failing which they will be referred to the JRC, and failing resolution at that level, the matter will be referred to the two Governments.

---

24 In the period 21–30 April when Bangladesh would have received 34,500 cusec under the old agreements, it now gets 35,000 cusec under the new Treaty, thus achieving a slight improvement—an important negotiating point on that country's part. On the other hand, it had been a grievance of West Bengal that the old agreements had been generous to Bangladesh at its cost; this is partly met by the allocation of 35,000 cusec to India in three 10 day periods out of six. It will be noted that this is really the equal sharing principle in a different form.

The principles of 'fairness, equity, and no harm to either side' are mentioned three times in the Treaty. There are also some declarations of good intention: a recognition of the need to cooperate in studying 'augmentation' possibilities, an agreement to conclude water-sharing treaties or agreements on other common rivers, and so on.

Reception of the Treaty

The conclusion of the Treaty was generally welcomed in both countries. Criticisms of the Treaty by some people in either country as a 'sell-out' to the other side tended to cancel each other out. In Bangladesh, the Bangladesh Nationalist Party (BNP), as the leading Opposition party, was understandably critical of the Treaty, and its declared position was that as and when it came to power, it would seek a revision of the Treaty. However, they did not succeed in mounting a national campaign against the Treaty. In India, there was some criticism on expected lines in West Bengal but it was fairly muted. At the national level the Bharatiya Janata Party (BJP) questioned the Treaty but this seemed a *pro forma* move rather than a serious criticism based on a careful analysis. Subsequently the BJP came to power at the Centre, but their Government has been scrupulously adhering to the Treaty. There were complaints in Bihar and Uttar Pradesh that their needs had not been kept in mind in the Treaty. Their grievance arose partly from the fact that only the Chief Minister of West Bengal was associated with the negotiations, but there was also some uneasiness stemming from the fact that the Treaty required the Government of India to make every effort to protect the flows reaching Farrakka. (Considering the wasteful use of water in this subcontinent, particularly in irrigation, there is enormous scope—indeed an inescapable need—for better water management. If economy in water-use and conservation of this scarce resource are given the importance they deserve, both India and Bangladesh can do with much less water. Neither Bihar nor Uttar Pradesh need have any apprehensions; but they may need to be given appropriate reassurances and assistance by the Centre.)

The Ganga Treaty in Operation

The Treaty has now completed six lean seasons. The flows in the river were very good in the 1998 lean season and they were above average even in the 1999 season. No problems appear to have been encountered in the 2000, 2001 and 2002 seasons also. The Treaty has operated smoothly, and Bangladesh has been receiving its share of the waters and even more. A few years ago, the India-Bangladesh JRC had expressed satisfaction with the working of the Treaty.

The 1997 Problems

However, during the very first lean season following immediately after the signing of the Treaty (January 1997 to May 1997), the Treaty had run into some rough weather, and it is necessary to go into this as it could happen again.

Low flows: The most important problem was the occurrence of low flows soon after the Treaty went into operation. This was a normal hydrological phenomenon; it has happened before and will happen again. Nor was 1997 a particularly bad year: there had been past years in which the flows had fallen to lower levels and for longer periods[25]. However, the unfortunate occurrence of low flows in the first lean season was interpreted by many in Bangladesh as evidence of a failure of the Treaty or of deficiencies in it. This was not true. In the event of the flows falling below 50,000 cusec (as they did in the last 10 days of March 1997), Article II(iii) of the Treaty provides for immediate consultations on an emergency basis. The consultations were initiated but not pursued to any useful result for various reasons: no formula was put in place for the equitable sharing of low flows.

Alternate 10-days Pattern: The pattern of sharing (35,000 cusec to either side in alternate 10-day periods between 11 March and 10 May) means there is a sharp variation in flows from one 10-day period to the next. There is a safety problem involved in this political solution, particularly when the flows reaching Farakka are low: a drastic drop in the water level in the feeder canal on the Indian side, if carried out rapidly, could lead to a collapse of the canal sides. Engineers on both sides seem to accept this. In the absence of an understanding between the two

---

25 It may be added that there is no evidence so far that increased upstream uses have led to a significant reduction in the flows at Farakka, compared to the flows last jointly determined in the 1970s. As the flows were very good in subsequent years, it is clear that 1997 was merely a relatively bad year, and did not indicate a trend.

countries on this subject, there was an unfortunate muddle in 1997. A gradual reduction in the canal level in the interest of safety resulted in a shortfall in the releases to Bangladesh in the first 10 days of April (compensated within the next 10-day period), and led to charges of violations of the Treaty. There is clearly a need for an early intergovernmental understanding on this matter. (It is possible that some working arrangement has been agreed upon, as one did not hear references to this problem in subsequent years).

Farakka–Hardinge Bridge Discrepancy: The third problem was a puzzling discrepancy (quite substantial) between the quantum of water released at Farakka in India and the quantum arriving at Hardinge Bridge (170 kilometres downstream) in Bangladesh. If the observations at Farakka and Hardinge Bridge were taken as reasonably reliable (as they are jointly monitored at both places), and if there were no significant abstraction of water in between (as both sides seemed to agree), then it appeared possible that the river water was finding its way into underground aquifers in Bangladesh. This is merely a hypothesis, and there may be other factors at work. Similar discrepancies had occasionally been noticed in the past too. Quite possibly, the problem was accentuated in 1997 because of low flows. An expert team was set up by the Joint Committee to examine this matter. Here again, one has not heard anything more on the subject.

Gorai problem: The offtake point of the Gorai river in Bangladesh has been silting up, and over the years a massive barrier has come up which prevents the entry of the waters of the Ganges into this stream, leading to problems downstream (loss of irrigation, salinity incursion, etc.,). The problem ante-dates Farakka. The Treaty is not going to solve this problem, because even 35,000 cusec will not be enough for the purpose; it is only when the flow is of the order of 70,000 cusec that the waters of the Ganges can enter the Gorai. The answer to this problem is perhaps partly extensive dredging and partly arrangements to head up the Ganges waters and enable them to enter the Gorai. (Some action on these lines is under the active consideration of the Government of Bangladesh).

Bangladeshi Perceptions: On an overall view, the position would seem to be that the Treaty had worked reasonably well even in the first lean season despite low flows, except for a very brief slip-up. However, perceptions in Bangladesh were quite different. Several factors combined to produce this result.

First, what was actually received by Bangladesh had to be compared with what it was entitled to under the Treaty in terms of the Annexure I formula (i.e., *a proportion of actual flows*); however, the comparison was often with *a proportion of the 40-year-average flows given by way of illustration in Annexure II*, treating these as 'expected' flows, and this fallacious comparison was bound to give rise to wrong conclusions.

Secondly, the shortfall in Bangladesh's share in the first 10-day period of April 1997 (which ought not to have happened) arose out of a concern for safety, and a prompt attempt was made to compensate for the short-fall.[26] This was not known to the people either in India or in Bangladesh.

Thirdly, the fact that the discrepancy between releases at Farakka and arrivals at Hardinge Bridge was a complex matter needing a scientific study had not been explained to the people; they were therefore ready to be persuaded by the opponents of the Treaty that they were being short-changed at Farakka.

Fourthly, the fact that the Gorai offtake problem was essentially an internal one needing remedial action in Bangladesh was not widely known. The general tendency was to assume that it had been caused by Farakka and to be disappointed that the Treaty had not had much impact on the problem.

Those four problems/factors tended to sow the seeds of suspicion; and the seeds fell on fertile ground. Because of long-standing emotional attitudes towards India, there was a predisposition on the part of many to view the Treaty negatively. As soon as problems emerged, the latent negativism came to the fore: the immediate and visceral tendency was to find explanations in terms of Indian wrongdoing. This was rendered easier by the absence of official information about the operation of the Treaty. The greatest enemy of the Treaty (and of India's own fair image) was the classification of information, i.e., the treatment of all informa-tion regarding flows and releases as secret.

While good flows in the river and the smooth operation of the Treaty in subsequent years have pushed the strains of 1997 into the background,

---

26 In fact, twice during the lean period India moderated its own withdrawals to ensure better flows to Bangladesh, and in the period as a whole Bangladesh received in the aggregate a slightly larger quantum of water than its share. This does not justify a shortfall in one particular 10-day period, but it is at least an indication of good faith and honourable intentions on the part of India.

they can come to the fore once again if the phenomenon of low flows were to recur, as it can. A good problem-free period provides an opportunity for the two Governments to reach an understanding on a contingency plan for a difficult time; but alas, they will probably do nothing until the difficulties are once again upon them: that is the nature of politics.

(Despite earlier criticisms of the Treaty by the BNP, that Party, when it came to power in Dhaka, did not ask for a revision of the Treaty. In fact, a Minister in the BNP Government appears to have acknowledged that the Treaty has been useful to Bangladesh. The Treaty provides for a review at the end of five years, and that point was reached in December 2001. However, neither the Government of India nor that of Bangladesh appears to have moved for a review. Realism and good sense seem to have prevailed on both sides: there is clearly a reluctance to disturb something that is working reasonably well.)

Concluding Reflections

Indus Commission and Joint Rivers Commission

A question that is sometimes asked is: why is it that the Indus Commission has been working reasonably well whereas the India–Bangladesh JRC has not been a great success? The answer is twofold. The first part is that the Indus Commission came into being under the Treaty that settled the water dispute; it merely had to monitor the implementation of the Treaty. On the other hand, the JRC had to function in a situation of a bitter unresolved inter-country dispute, and became a part of the negotiation mechanisms. The second part is that even after the signing of the Ganga Treaty, the JRC has a vastly more complicated and difficult task to perform than the Indus Commission. As we have seen, the Indus Treaty finally allocates some rivers as a whole to Pakistan and others to India, and does not allocate shares in the same river. Thus, there is no continuing process of water-sharing in a given river, requiring operations, measurements, monitoring, etc., as is the case under the Ganga Treaty. Further, the JRC is concerned with all the rivers common to India and Bangladesh, the Ganga being only one of them and one that is covered by a Treaty.

Water Disputes and Politics

There is a complex interaction between water issues and political relations. It is not always a case of conflicts over water resources leading to
a worsening of political relations, though that does happen on occasion;
it is more often a case of a difficult political relationship rendering the
water issue more intractable. This is particularly so when other issues
become prominent from time to time: for instance, Nepal's grievances
as a landlocked country and its concerns over trade and transit issues;
or India's security concerns and its apprehensions about Nepal's relations with China; or in the Indo–Bangladesh context, the question of
illegal immigrants, Chakma refugees, insurgency operations, border
demarcation issues, trade balance, etc. Water issues in turn can become
the most dominant factor at certain times, and can have a decisive impact
on the general political relationship.

Big Country/Small Country

More than anything else, one factor tends to influence Indo–Nepal and
Indo–Bangladesh relations, namely, India's size. India's relations with
its smaller neighbours have doubtless improved significantly in recent
years, but the old distrust of the 'Big Brother' has not wholly disappeared
either in Nepal or in Bangladesh. The twin dangers of big-country insensitivity or arrogance and small-country pathology are the Scylla and
Charybdis that can wreck even a good relationship. Everyone recognizes
the first danger; but not many perceive or understand the second.

There is no doubt that Indian politicians, bureaucrats, engineers and
businessmen have on occasion been unimaginative, patronizing and in-
sensitive in their dealings with the country's smaller neighbours; and
there have been brief aberrant periods when even the word 'bullying'
might not have been out of place. However, underlying all this there has
always been a deep desire in India, at the political, administrative and
popular levels, to be liked and approved by the rest of the world, and
in particular by the neighbouring countries; and this cuts across party
affiliations and other dividing lines. There is also a fairly widespread
acceptance (with exceptions, of course) of the proposition that India,
as the bigger country, must go more than halfway in seeking to build
good relations with the smaller neighbours. This was so even before the
'Gujral doctrine' of unilateral gestures without asking for reciprocity was

formulated. Indians like to believe that they are good, friendly, fair-minded fellows; that is why accusations of bad faith, bullying and hegemonism evoke reactions of shocked surprise, self-righteous indignation and hurt feelings on the part of India. There is a failure to recognize the policy errors and behavioural insensitivities that lead to such accusations.

The counterpart to big-country insensitivity, namely, the 'small-country pathology' referred to above, manifests itself in many ways: a 'tough' stance during negotiations for fear of being considered weak; complaints at a later stage that the negotiations had been between un-equal parties; when difficulties or differences emerge in the course of operation of a Treaty or agreement, a tendency to seek explanations in terms of deviousness or machinations or malevolence on the part of India instead of exploring solutions, and so on. Immediately after a Treaty is signed, the refrain begins: 'It is not enough to sign a Treaty, it must be properly implemented', implying doubts about India's good faith. Whenever differences arise, the tendency is not to recognize it as a differ-ence needing to be mutually resolved but to regard it as a case of 'non-implementation' or 'violation' of the Treaty. All this is very understandable, and India will have to learn to live with this, find constructive solutions to problems and at the same time maintain and improve relations with its smaller neighbours; but the phenomenon mentioned does exist and needs to be recognized.

Riparian Rights

India's position as the middle riparian (between Nepal and Bangladesh), together with its desire to avoid accusations of big-country dominance, places it in a somewhat difficult situation. For instance, *vis-à-vis* Nepal, India as the lower riparian has conceded to the upper riparian the right to use whatever waters the latter needs, including the right to make inter-basin transfers. Thus, primacy has been given to Nepal, and the lower riparian's rights have been subordinated to those of the upper riparian. On the other hand, *vis-à-vis* Bangladesh, it is the primacy of the lower riparian that has been implicitly (partly) granted. The whole complaint of Bangladesh for years has been that the flows coming down have been reduced by the upper riparian; and they continue to express apprehensions about upstream uses in Uttar Pradesh and Bihar, almost as if these were not legitimate. To meet the concerns of Bangladesh, India as the upper riparian has undertaken to protect the flows reaching Farakka, which means that it has undertaken the responsibility of regulating upstream uses. It is not being argued here that India was wrong in accommodating Nepal's and Bangladesh's points of view on the lines mentioned above; the point is merely that there is in fact a certain asymmetry in the Indian position *vis-à-vis* the two countries, arising from India's size and its neighbours' concerns, and that this must be noted.[27]

The 'No Harm' Principle

In both the Mahakali and Ganga Treaties the principle of 'no harm to either side' has been incorporated. That seems reasonable, but it could present some difficulties at a future date. It is fair enough to say that the upper riparian, in planning its own water-uses, should keep the interests of the lower riparian in mind and refrain from causing undue injury or distress to it; but the absolute assertion of 'no harm' could put the upper riparian in some difficulty, as any use by it of the waters of the river in question may conceivably reduce downstream flows, and therefore cause some harm, however minor, to the lower riparian. The Helsinki Rules wanted the upper riparian to avoid causing substantial harm, and the present UN Convention uses the expression 'significant' adverse effects. In the Ganga Treaty there is no such qualification. In practical terms, this may not cause any difficulty in the near future, but the potential for a difference in perceptions at some stage does exist. This will need to be avoided by wisdom and goodwill on both sides.

Bilateralism versus Regionalism

An issue that needs to be taken note of here is that of 'bilateralism' versus 'regionalism'. The Government of India has in the past preferred to deal with Nepal and Bangladesh separately and bilaterally, and though this

---

27 It may be added that some in Bangladesh go so far as to argue that it is a 'stakeholder' in the entire Ganga system and should be consulted or kept informed about any intervention, major or minor, at any point in the system from the Gangotri downwards. It is not possible to go into this in detail here; it is mentioned in passing only as an indication of an extreme lower riparian position.

approach is getting affected by India's participation in the SAARC, the official policy still remains one of bilateralism. The considerations behind this position seem to include: (*a*) the feeling that negotiations are difficult and protracted enough between two countries, and that they will become vastly more complicated and intractable when three or more countries are involved; and (*b*) the fear that the smaller countries may join hands and make common cause against the bigger country. These are exaggerated fears, though they cannot be dismissed as totally unfounded and absurd. On the other hand, a view that has gained considerable currency, not only in Nepal and Bangladesh but even among many in India, is that 'regional cooperation' has better solutions to offer to most problems than purely national or bilateral approaches. On this, as on many other subjects, pragmatism is preferable to dogmatism. We need subscribe neither to the rigid and myopic bilateralism of the Government of India, nor to the doctrinaire regionalism that is much in fashion now. Some issues are essentially bilateral. For instance, reaching an understanding on the sharing of Ganga waters as they cross the Indo–Bangladesh border is clearly a bilateral issue; and the Treaty would never have come about if it had become the subject of multi-country discussions. Similarly, the Tipaimukh Project near the Indo-Bangladesh border in Tripura (if it ever comes into being) would concern only India and Bangladesh. On the other hand, the Kosi High Dam Project (again, if it ever gets approved) is of interest to Nepal, India and Bangladesh, and could well be conceived of as a joint three-country project.

Leaving particular projects aside, it could be plausibly argued that there should be 'integrated' planning for a basin as a whole, because that is the hydrological unit, and that this calls for a regional approach. Theoretically this may be true, but in practice there are serious difficulties. First (as we have seen), there is no agreement even on what constitutes a basin: is Ganga–Brahmaputra–Meghna one basin or three basins? Behind the difference on that question between India and Bangladesh lie not hydrological but other considerations: India opts for a 'single basin' theory because it is anxious to tap the waters of the Brahmaputra and take them westwards; Bangladesh insists that they are three basins because it wants to maintain a legal claim on the Ganga as a separate river and feels that this may be compromised if the Ganga were regarded as a part of a larger system. Secondly, the idea of planning for a basin or sub-basin as a whole has not made much headway even within India: it has not so far been possible to get State Governments (e.g., Tamil Nadu and Karnataka; Andhra Pradesh and Karnataka) to agree to such an approach to the resolution of inter-State disputes. A similar approach involving several countries is an enormously more difficult and complex undertaking. The Mekong Commission is a good example in so far as countries that had earlier gone to war are now sitting together and talking across a table, but in terms of achievements it has (so far) not much to show. Thirdly, the commitment of Nepal and Bangladesh to the idea of basin-wide planning is imperfect. On the one hand the Mahakali Treaty is described as a Treaty concerning the integrated development of the Mahakali river; on the other, Nepal argues that the Treaty is only about the river in the border stretch. Similarly, Bangladesh talks about an integrated approach but insists on a river-by-river sharing. Doubtless these positions of Nepal and Bangladesh are derived from certain concerns that India must take into account. The point being made here is merely that neither country subscribes unreservedly to the 'integrated', 'regional' approach that it advocates. These difficulties are not arguments against either regional cooperation or basin-wide planning. These may be goals towards which we must start moving now, but this need not prevent more modest achievements in cooperation in the nearer future. The caution here is against: (a) a doctrinaire approach, and (b) a uniform prescription of regionalism as appropriate to all matters under discussion, and a dismissal of a bilateral approach as inherently inferior.

Cooperation: Wider Meaning

Further, as mentioned in Chapter 6, the language of 'integrated, basin-wide planning' seems to carry implications of centralized technology-driven planning; and 'regional cooperation' usually implies cooperation at the governmental or technocratic levels. Both these terms need to be interpreted in a wider sense so as to cover cooperation at levels other than that of governments, and for purposes beyond engineering or technological ones.

It is often argued that in our water resource planning we must take note of the immense possibilities offered by regional cooperation, particularly cooperation with Nepal and Bangladesh, on the utilization of the water resources of the Ganga, Brahmaputra and Meghna rivers. While wholly endorsing the plea for cooperation, one still needs to ask: in concrete terms, what does 'cooperation' mean here? It means essentially cooperation in 'harnessing' the water resources of the GBM system(s) by means of a number of major projects. However, these will need giant dams and reservoirs in a fragile and seismically active ecosystem. Whether, and if so when, any or all of these projects will in fact come up is a question that no one can answer with any confidence. Meanwhile there are many other possibilities and compulsions of cooperation. Relatively smaller, less conflict-prone and more easily manageable projects; the protection of water sources (rivers, lakes, mountains, forests, aquifers) from pollution, degradation or denudation; the preservation and regeneration of deteriorating wetlands (e.g., the Sunderbans); improving and maintaining water quality; dealing with common problems such as drainage in the Indus basin in both India and Pakistan, or the occurrence of arsenic in aquifers in both India and Bangladesh; coping with floods and minimizing damage; sharing experiences in local water-harvesting and watershed development, and in the related social mobilization and transformation: these are among the areas in which inter-country cooperation will be very fruitful, and in some instances very necessary. Such cooperation can be at the level of governments, NGOs, academic institutions, 'think tanks' or 'people-to-people'. Such possibilities have not received the attention they should have.

20

The Fallacy of 'Augmentation'

General Agreement on 'Augmentation'

In the Indo–Bangladesh talks over the past two decades and more regarding the sharing of the waters of the Ganga at Farakka, a persistent theme has been 'augmentation'. As early as 1974, it was agreed in the 'Joint Declaration of the Prime Ministers of India and Bangladesh' that there was not enough water in the Ganga in the lean season to meet the combined needs of the two countries, and that those flows would need to be augmented. The 1977 Agreement between the two countries was both a 5-year water-sharing agreement and an agreement to work out an acceptable proposal for augmenting the lean-season flows. 'Sharing' and 'augmentation' were two equally important and linked components of that Agreement, as also of the 1982 and 1985 Memoranda of Understanding. It was only the 1996 Treaty that for the first time refrained from linking the two ideas: it was essentially a water-sharing Treaty. Nevertheless, that Treaty too contained a reference to augmentation in the Preamble as well as in Article VIII, which stated that 'the two governments recognize the need to cooperate with each other in finding a solution to the long-term problem of augmenting the flows of the Ganga/Ganges during the dry season'.

Thus 'augmentation' is an old and accepted idea. There was a divergence between India and Bangladesh on how the augmentation was to be achieved. The Indian view was that the Ganga was water-short and needed to be augmented by a diversion of waters from the Brahmaputra which had a large surplus. Bangladesh argued for an augmentation within the Ganga system by storing a part of the seasonal high flows behind seven large dams in Nepal. Each side had serious reservations about the other's proposal. This divergence was one of the reasons for the prolonged impasse between the two countries over the Ganga waters issue;

eventually, it was only by putting the augmentation question aside that a sharing Treaty could be concluded. It is not necessary to go into that old controversy here. What needs to be noted is that both sides are in complete agreement on the *need* for an augmentation of the lean season flows of the Ganga.

A Dubious Proposition

The case for augmentation can be broken down into two propositions: (i) that there is a shortage of water in the Ganga, and (ii) that this should be made good by an addition of water to the Ganga from somewhere. Both those propositions need to be looked at closely, but first the very notion of 'augmentation' (which has become so familiar that we tend to take it for granted) has to be examined.

'Augmentation' implies an *addition* to the available water. But in what sense can we *add* to the quantum of water in the Ganga? We cannot increase what is available in nature. What we *can* do is to hold back a part of the high flows behind a dam and release it during the lean season. In other words, 'augmentation' means no more than a modification of the seasonality of the flows. This can be thought of not only in the case of the Ganga, but in the case of other seasonal rivers as well. Should we then undertake the modification of the seasonality of all seasonal rivers? Such a proposition is not likely to be seriously argued. There is no such thing as an augmentation (i.e., seasonality-modification) project: no project is ever proposed merely for that purpose. If dams are to be built at all, they will have to be built for other purposes, such as irrigation, power-generation, flood-moderation and so on. The 'augmentation', if any, of lean-season downstream flows can only be an incidental and secondary consequence of a project built for other purposes.

Further, while *augmentation* (of lean-season downstream flows) is problematic because it is bound to be subordinated to the primary objectives of the project in question, a *reduction* of downstream flows in the major part of the year is a certainty. All dams stop flowing waters for the purpose of storing, diverting and using them, and must therefore necessarily reduce downstream flows. This usually figures as a prominent element in the enumeration of the adverse impacts of damming a river. A reduction of downstream flows could have several consequences: it could affect the river regime, cause serious problems to aquatic life and riverside communities, hamper economic activities such as fishing, plying of boats, agriculture, etc., impair the self-regenerating capacity of the river and affect water quality, diminish groundwater recharge, adversely affect estuarine conditions, lead to the increased incursion of salinity from the sea, and so on. One of the issues in the Narmada (Sardar Sarovar Project) controversy is the impact of the dam on downstream areas. In the Cauvery dispute, the complaint of Tamil Nadu is that the flows into that State have been reduced by the building of storages by Karnataka. Lower riparians can be normally expected to object to the building of dams upstream. It is therefore rather strange that Bangladesh is actually asking for dams to be built in Nepal! If Farakka had not come into existence and the question of 'sharing' and 'augmentation' had not arisen, Bangladesh would probably have strenuously objected to any dam-building proposals in India or in Nepal on the ground of adverse impacts. The short point is that 'augmentation' and 'reduction of flows' are the obverse and reverse of the same coin, and that we cannot have the one without the other.

A Digression regarding the Brahmaputra

A diversion of waters from the Brahmaputra could of course mean an addition to the flows of the Ganga. Bangladesh is not convinced that there is in fact a surplus in the Brahmaputra for diversion; leaving that aside, we must note that such a diversion project would have its own environmental, social and human consequences. The Indian proposal (of the 1970s) of a Brahmaputra–Ganga link canal running through Bangladesh was objected to by Bangladesh on many grounds, at least some of which were valid. That proposal is not being seriously pursued now, and in any case, it is very unlikely that Bangladesh will ever accept it. The alternative of a link canal running entirely through Indian territory seems unlikely to be a viable proposition. Such long-distance water-transfers are highly problematic and extremely difficult to justify. In so far as India wishes to utilize the waters of the Brahmaputra, it will have to explore more modest possibilities within the North-east and the immediate neighbourhood (perhaps including limited transfers between adjoining rivers within the Brahmaputra system), rather than think in terms of a large water-transfer westwards.

The 'Shortage in the Ganga' Argument

'Augmentation', then, is a very dubious proposition. Let us now return to the question of the *need* for it, as perceived by the two Governments from the 1970s onwards. Taking the 'shortage' argument first, the perceived inadequacy of Ganga waters is based on the fact that while the lowest '75 per cent dependable' flow in the river in the leanest part of the lean season was determined as 55,000 cusec, each of the two countries projected its own need for Ganga waters as equal to or exceeding that quantum. However, how reliable are those estimates of demand? Apart from the fact that each country tended to overstate its claim because of the existence of a dispute, it must also be noted that the projections were based on the prevailing patterns of water-use. If a careful estimate were made in either country with due regard to economy and efficiency in the use of this precious natural resource, we may well find that the two countries can manage with much less water than they think they need. The 'shortage' in the Ganga may turn out to be a myth based on the present wasteful and profligate use of water. In terms of water endowment, the areas that we are concerned with here are not among the most water-stressed in the world; there are countries (in the Middle East, for instance) that have much less water and have learnt to use that water very carefully. From their point of view, our complaints about a shortage of water in the Ganga may sound strange and incomprehensible.

Put More Water into the Ganga?

As for the second part of the proposition, assuming that after a careful examination the combined needs of the two countries are in fact found to exceed the lean-season flows in the Ganga, does it follow that those flows should be 'augmented', i.e., that more water should be added to the Ganga from somewhere, whether from within the Ganga system or from another river, for being taken out and shared? That is a very odd idea, the strangeness of which has been blunted by repeated iteration. It is necessary to get away from familiar and accustomed grooves of thinking and take a fresh look at this matter.

Where a river runs through more than one country, the countries concerned must reach an agreement on the sharing of the waters of that river. Such an understanding has in fact been reached in relation to the Ganga, and embodied in the 1996 Treaty. There is nothing more to be done in so far as the Ganga is concerned, except to resolve such practical difficulties as arise in the actual operation of the Treaty. If indeed there is not enough water in the Ganga, then the two countries should look to other sources. There is no reason why they should think in terms of putting more water into the Ganga ('augmenting' its flows) for the purpose of being taken out and shared, as if all water must necessarily come only from the Ganga.

Consider Totality of Resource

What the two countries must do is to take a look at the totality of their water needs and the totality of water availability in all forms (other rivers, surface-water bodies, groundwater, deep aquifers, soil moisture, wet-lands, and of course rainfall), and make careful plans for proper resource management on a holistic and integrated basis. (Incidentally, local water-conservation through water harvesting and watershed development would in fact constitute true 'augmentation').

Better Resource Management

The recommendation that emerges from the foregoing analysis is that the two countries should stop talking about the 'augmentation' of the lean-season flows of the Ganga, and instead, think in terms of proper water resource management. Such efforts will be essentially internal to each country, but there can be a sharing of experiences between countries, and cooperation in areas close to one another.

It is conceivable that one or more of the projects in Nepal (Karnali, Pancheswar, Sapta Kosi) that have been talked about may survive the most stringent scrutiny from all angles (including those of seismicity and downstream impacts), prevail over all opposition, obtain the necessary finances and actually get constructed; and that it (or they) may in fact result in increased lean-season flows at Farakka as a secondary consequence. One is profoundly sceptical regarding this possibility, but if it happens, the increased flows will of course be available for being shared by the two countries under the Ganga Treaty. This chapter is questioning not such a contingent (but remote) possibility, but the soundness of 'augmentation' as an idea and an objective.

VI

Water: Looking at the Future

21

Water: The Basics

Preliminary

In recent years there has been a growing perception of a looming water scarcity. Several institutions and networks sprang up to deal with this and related matters, and many exercises were undertaken during the years 1997 to 2000 in preparation for the massive World Water Forum held at The Hague in March 2000. Now work is going on in preparation for the next World Water Forum to be held in Kyoto in 2003.

A common trend in most of the discussions (those preceding the Hague Forum as well as those at the Forum sessions) was to proceed from projections of demand to supply-side solutions in the form of 'water resource development' projects; estimate the massive investment funds needed; take note of the severe limitations on the availability of financial resources with governments; point to private sector investments as the answer; and stress the need for policy changes to facilitate this.

Within India, a consciousness of the importance of the subject led to the establishment of a National Commission for Integrated Water Resources Development Plan (NCIWRDP) in 1996, and the Commission—the first National Commission on water—submitted its Report in September 1999. The Report covered extensive ground and made numerous recommendations, but unfortunately it has not been widely disseminated.

At this stage, before proceeding to a consideration of future courses of action, some basic information may be useful for the general reader, even if this covers territory familiar to more knowledgeable people. (The author's source for much of the information in the following paragraphs was the material that became available to him as a member of the NCIWRDP.)

Some Fundamentals

Superficially, water seems over-abundant on this planet: three-quarters of its area is covered by water. The 1,400 million cubic kilometres (km^3) of water so present can cover the entire area of the earth to a depth of 3,000 meters. However, around 98 per cent of the water is in the oceans. Only 2.7 per cent is fresh water; of this 75 per cent lies frozen in the polar regions; 22.6 per cent is present as groundwater, some of which lies too deep; only a small fraction is to be found in rivers, lakes, atmosphere, soil, vegetation and exploitable underground aquifers, and this is what constitutes the fresh water resources of the world. Annually, 300,000 km^3 of precipitation takes place over the oceans, and 100,000 km^3 over land; evapo-transpiration from land is 60,000 km^3, 40,000 km^3 run off from land to sea and 340,000 km^3 evaporate from the seas. This is the annual hydrological cycle.

In this context, two points of a fundamental nature need to be kept in mind. The first is that water in all its forms (snow, rain, soil moisture, glaciers, rivers, lakes, other surface-water bodies and groundwater) constitutes a unity. The second is that there is a finite quantity of water on earth, and this is neither added to nor destroyed.

We cannot create new water, and whatever quantity is used up in any manner substantially reappears, though not always or fully in a reusable form. A significant part of the water applied to fields in irrigation either seeps through to underground aquifers, or reappears as 'return flow' and finds its way back to the surface (this is sometimes described as 'regeneration'); seepages from canals recharge groundwater aquifers; industrial use of water results in effluents; domestic and municipal uses become sewage, and of course, whatever water evaporates comes back to earth as rain or snow.

The water available to us on earth today is no different in quantity from what was available thousands of years ago. That finite quantity has to be juxtaposed against increasing demands from a growing population. The population of the world, currently around 6 billion, is expected to exceed 8 billion by the year 2050. Apart from sheer numbers, the processes of urbanization and 'development' are also expected to result in a vast increase in the demand for fresh water. It is this factor that leads to projections of water scarcity; and the scarcity could be severe in some parts of the world.

However, while all this may be useful by way of background information, global figures are not of much practical significance. Water is not an internationally traded commodity to an extent even remotely comparable to oil, and the availability of water in a distant part of the world is of no great relevance to a water-short country or region. People need sources of water close to their homes and lands.

India: Some Facts

With a population that is 16 per cent of the world's, India has 2.45 per cent of the world's land resources and 4 per cent of its water resources. The average annual precipitation by way of rain and snow over India's landmass is 4,000 km^3, but the 'available' water resources of the country are measured in terms of the annual flows in the river systems. This has been estimated by the NCIWRDP as 1,953 km^3. Some of the water resources of the country flow into it from beyond our borders—say, from Nepal or Tibet—and some cross our borders and go into other countries (Pakistan, Bangladesh). We have expectations of flows from the 'upper' countries and obligations to the 'lower' countries.

Turning to groundwater, the quantity that can be extracted annually, having regard to the rate of annual replenishment (recharge) and economic considerations, is known as the 'groundwater potential'. This has been put at 432 km^3. Extraction exceeding the rate of recharge is known as 'mining'.

Whether those two figures, namely, 1,953 km^3 of river-flows and 432 km^3 of groundwater, can be regarded as distinct, or whether some double counting is involved in this because of the inter-relationship between surface water and groundwater, is a debatable proposition.

Only a part of the 'available' water resources is considered actually 'usable' in practical terms. The National Water Policy (NWP) 1987 had a proposition urging that more of the 'available' quantum of water be brought into the 'usable' category. The NWP 1987 did not elaborate that statement, but it was generally understood in the water resources establishment as referring to 'water resources development' or 'WRD' projects. (The NWP 2002 spells out the notion in a somewhat wider form.)

It has been estimated by the NCIWRDP that the annual 'usable' water resources of the country are 690 km^3 of surface water and 396 km^3 of groundwater, making a total of 1,086 km^3. The present quantum of use is put at around 600 km³. It follows that in national terms the position is not uncomfortable at the moment. However, this will obviously change with the growth of population and the processes of urbanization and 'development'. The NCIWRDP has made various assumptions in regard to these matters (high, medium and low rates of change), and come to the conclusion that by the year 2050 the total water requirement of the country will be 973 to 1,180 km³ under 'low' and 'high' demand projections, which means that supply will barely match demand. It is the Commission's view that there will be a difficult situation but no crisis, *provided* that a number of measures on both the demand side and the supply side are taken in time. (The precarious balance between supply and demand can of course tip over into a crisis if the actual developments fail to conform to the assumptions. Moreover, apart from demand putting pressure on the available supplies, the supplies themselves may also be seriously affected by the growing incidence of pollution and contamination of water sources.)

A word regarding the concept of 'water stress' may not be out of place here. Dr Malin Falkenmark, the leading Swedish expert, has calculated the 'water stress' situation of different countries with reference to 'annual water resources per capita' (AWR). An AWR of 1,700 m³ means that only occasional and local stress may be experienced; an AWR of less than 1,000 m³ indicates a condition of stress, and one of 500 m³ or less means a serious constraint and a threat to life. Under this categorization, India is somewhere between categories (i) and (ii). In other words, India is not among the most water-stressed countries of the world. Israel, for instance, has a much lower endowment. But this situation will change with the growth of population, and India may join the ranks of 'water-stressed' countries in the future if counter-measures are not taken.

Variations

However, national aggregates and averages are as misleading as global figures. There are wide variations, both temporal and spatial, in the availability of water in the country. Much of the rainfall occurs within a period of a few months during the year, and even during that period the intensity is concentrated within a few weeks. Spatially, there is a wide range in precipitation—from less than 200 millimetres (mm) (or even 100 mm) in parts of Rajasthan to 11,000 mm in Cherrapunji. (Incidentally, it must be noted that despite the very heavy precipitation,

Cherrapunji, known as among the wettest places on earth, suffers from an acute shortage of water during some parts of the year, because the rainfall quickly runs off the area.) Sixty per cent of the water resources of India are to be found in the Ganga, Brahmaputra and Meghna river systems which account for 33 per cent of the geographical area of the country; 11 per cent in the west-flowing rivers south of Tapi covering 3 per cent of the area; and the balance 29 per cent in the remaining river systems spread over 64 per cent of the land area. Broadly speaking, the Himalayan rivers are snow-fed and perennial, whereas the peninsular rivers are dependent on the monsoons and therefore seasonal; and again broadly speaking, the north and east are well endowed with water whereas the west and south are water-short. Apart from the desert areas of Rajasthan, there are arid or drought-prone areas in parts of Gujarat, Maharashtra, Karnataka, Andhra Pradesh and Tamil Nadu; and of course the eastern parts of the country experience devastating floods from time to time.

The Standard Response

The standard engineering response to these temporal and spatial variations is to propose: (a) the storing of river waters in reservoirs behind large dams to transfer water from the season of abundance to that of scarcity, and (b) long distance water-transfers from 'surplus' areas to water-short areas. To projected future demands, supply-side solutions in the form of large dam-and-reservoir projects are believed to be the proper answer; and for water-scarce areas, the answer is believed to lie in bringing in water from distant areas. Both large 'storages' (i.e., reservoirs) and the 'linking of rivers' (i.e., inter-basin transfers) have played an important part in the thinking of our water resource planners, and both involve major engineering interventions in the form of large projects.

(Apart from the spatial and temporal variability in the availability of water, another basic determinant in the modern engineering approach to water resources development in India is a view of rivers as 'surface-water resources' to be 'harnessed' and 'exploited' for human use with the instrumentality of science and technology. 'Use' is generally understood to mean 'out-of-stream' use, i.e., abstraction from the stream, largely for irrigation. There is a failure to recognize that 'in-stream' water also serves some important purposes; and there is a widely prevalent tendency to think of water that goes to the sea as 'wasted'.)

A major concern of our planners has been the consideration that a significant part of India's water resources is in the Brahmaputra and that ways and means must be found of 'harnessing' those resources and taking them westwards and southwards to areas that are water-short. This was the thinking behind the Indian proposal of the 1970s (in the context of the Indo–Bangladesh talks over Ganga waters) for a gravity link canal between the Brahmaputra and the Ganga through Bangladesh. That proposal was strongly objected to by Bangladesh and is no longer being seriously pursued, but the idea of tapping the waters of the Brahmaputra continues to exercise the minds of our water planners. Similarly, three decades after Dr K.L. Rao mooted the notion of a Ganga–Cauvery link, and Captain Dinshaw J. Dastur, a pilot, came up with the proposition of a 'Garland Canal', these ideas, long ago discarded as impracticable, continue to beguile the minds of the Indian public, particularly in the water-short south. For over two decades the National Water Development Agency (NWDA) has been studying the resources of different basins, assessing the availability of surpluses for transfer and identifying possibilities of storages, links and transfers. They took up the peninsular rivers first, and studied the possibility of transferring waters from the Mahanadi to the Godavari and thence to the Krishna, Pennar and the Cauvery, though it is difficult to persuade Orissa and Andhra Pradesh that there is a surplus in the Mahanadi and in the Godavari. Another idea that has been mooted is the diversion of west-flowing rivers eastwards, but there is resistance to this too. In recent years the NWDA has been studying the Himalayan rivers, but this is an even more difficult subject.

'Water Resource Development'

The NCIWRDP stresses demand management, economy in water-use, resource-conservation, etc., and also devotes a whole chapter to local water-harvesting and watershed development, but the thrust of the Report is on large water resource 'development' projects which are regarded as the primary answer to the future needs of a growing population. The report also discusses the financing of projects and the contributory role of private sector participation in the massive effort that is envisaged.

Thus, both at the regional/international level and at the national level there seems to be widespread agreement: (a) that to the projected water needs of the future an important (if not the major) part of the answer lies in 'water resource development' projects for storage and transfer over time and/or space; and (*b*) that considering the financial constraints and managerial limitations of governments, at least a part of that development will have to come from the private sector. There are many issues involved in these matters, and some of these have been discussed elsewhere in this book. The present chapter seeks to provide only a broad background as a preliminary to the chapters that follow.

A Parenthesis

Incidentally, even regarding the projected water crisis, there are two views. One—a widely held view—is that with a finite supply and a fast-growing demand a crisis is inevitable and not very far away. The other is that with proper demand management, economy in water-use and extensive local water-harvesting and watershed development a crisis can be averted. The latter view was put forward in a paper circulated by the Centre for Science and Environment at the Hague Forum.

22

Problems, Weaknesses, Failures

There is no doubt that the projects and schemes undertaken in the past ('major/medium' irrigation and multipurpose projects, minor irrigation schemes based on surface water and groundwater, etc.,) have contributed (along with other factors) to an increase in food production, added to hydropower capacity, provided water for domestic, municipal and industrial uses, and (to some extent) helped in flood-moderation. However, there have been many problems, weaknesses and failures, and before we consider any future course of action, we must take a clear look at the past: diagnosis must precede prescription. This chapter will therefore be necessarily concerned with negative aspects.

Drinking Water

The National Water Policy (NWP) 1987 assigned the highest priority to drinking water, but like most statements in the NWP, this remained a mere declaration on paper. Despite five decades of planning and more than a decade of 'Drinking Water Missions', there are large numbers of 'no source' villages (i.e., those with no identified source of safe drinking water). The curious fact is that targets for covering such 'no source' villages are repeatedly achieved, but the numbers grow larger rather than smaller. This must mean that some 'covered' villages are lapsing back into the uncovered category, and that newer villages are being added to this class.[28]

A significant aspect of the scarcity of water in rural areas is that the burden of fetching water from distant sources falls on women

---

28 Cf. The following extract from an internal paper of the Planning Commission made available to the author:

(including girl children), and yet women who are the providers and managers of water in the household have little voice in 'water resource planning' in this country.

As for urban areas, most large cities are chronically short of water. A few illustrations may suffice. Chennai has been waiting for water from the Krishna under the Telugu Ganga Project (for which it has contributed large sums of money), but the partial supplies that began belatedly appear to have stopped because of some difficulties. Efforts to revive the old, abandoned Veeranam project also seem to have run into difficulties. Bangalore is hoping for water from the distant Cauvery IV project. Delhi is repeatedly asking its neighbouring States for more water, and is waiting for the fruition of some major (and distant) projects (Tehri, Renuka).

It seems clear that ensuring access to safe drinking water to all has not been among the successes of our water planning.

Drought-Prone Areas

There are many arid zones and drought-prone areas in the country: for instance, in Rajasthan, Gujarat, Maharashtra, Karnataka, Andhra Pradesh and Tamil Nadu. Droughts are a recurring feature in these areas, cause much misery to human beings and livestock, and often result in large-scale migration. Unfortunately, there is no well thought-out strategy for 'drought-proofing' these areas. The planners and engineers, whether at the State level or at the Central level, seem to be pinning their hopes on vague notions of long-distance water-transfers. There has been no serious attempt to work out a series of area-specific answers by way of local conservation and augmentation to the maximum extent possible.

'In 1972, surveys revealed that out of a total of 5,80,000 revenue villages there were 1,50,000 drinking water "problem villages" in India. By 1980, some 94,000 villages were covered by the government and 56,000 were left uncovered. But the 1980 survey revealed that the number of "problem villages" was actually 2,31,000 and not merely 56,000. By 1985, all but 39,000 villages were covered, but the new survey revealed 1,61,722 "problem villages". Again, by 1994, they were all covered leaving only 70 uncovered villages but the 1994 survey revealed 1,40,975 problem habitations. This time the number included both revenue villages as well as hamlets. ...'

The severe drought experienced by India in the summer of the year 2000, which, incidentally, affected Pakistan as well, was not an indication of 'water insecurity', nor did it point to the need for big projects or long-distance water-transfers. Most of the comments, analyses and prescriptions by experts that appeared in the media, whether in India or in Pakistan, recognized that the drought conditions were the result of poor water management in the past, and that the answer lay in better resource-management in the future. Failure to harvest rainwater, excessive extraction of groundwater and failure to ensure the recharge of aquifers led to the water table falling sharply over the years, so that when a bad year came there were no groundwater reserves to draw upon. That is a broad description of what went wrong, though conditions may have differed from place to place. The correctness of that explanation is proved by the fact that in the same areas (parts of Rajasthan and Gujarat in India) lush green villages were to be found beside parched brown villages: the former had been practising water-harvesting and groundwater recharge for some years, the latter had not.

Floods

This is another area of an absence of policy and strategy, though the NWP 1987 had something to say on the subject, and before that, in the 1970s, there was an elaborate Report by the National Flood Commission or *Rashtriya Barh Ayog* (RBA). The numerous recommendations of the RBA remain largely unimplemented. Governments have tended to react spasmodically whenever floods occurred in disastrous form. The initial response to flood damage was to try to 'control' floods through structural means such as dams or embankments. It was found through experience that these efforts were ineffective or even harmful. For instance, large dams are not often planned with flood-moderation as a primary aim, and even where they are, the competing claims of irrigation and power-generation often override the flood-moderation function.[29]

---

29 The Damodar Valley Corporation (DVC) was planned for multiple functions (flood-moderation, power-generation, irrigation and the general development of the area), but the flood-moderation achieved was not of the planned order. The functions of the DVC were whittled down over the years. The DVC today is mainly a power-generating body, and much of that power is ironically enough thermal power.

Further, while dams may moderate flood flows to some extent under normal conditions, they may aggravate the position if (in the absence of a flood cushion) water has to be suddenly released in the interest of the safety of structures. As for embankments, they have to be repeatedly rebuilt at great cost; they may fail in the event of a major flood and cause greater difficulties; by jacketing the river and preventing it from spreading they may create new problems further down; by blocking drainage from the adjoining areas into the river they often lead to waterlogging and 'man-made floods' in the 'protected' villages, and they deprive farmers of the benefit of the deposit of silt by receding floods. Thus embankments have often proved a remedy worse than the disease, and there is a powerful people's movement in Bihar against them. On the other hand, while flood-moderation has been very modest, the extent of suffering, damage and economic loss caused by floods and the magnitude of government expenditure on 'relief' has been growing because of a number of factors.[30]

It is increasingly recognized that what we must learn to do is not so much to 'control' floods as to cope with them when they occur and minimize damage, partly through 'flood plain zoning' (i.e., regulation of settlement and activity in the natural flood plains of rivers) and partly through 'disaster-preparedness'. Unfortunately, 'flood plain zoning' has been found politically difficult. As for 'disaster-preparedness', the most important element in this is timely knowledge. Governments, local bodies and people need to know how soon a flood is likely to arrive, and what its magnitude is likely to be. They can then take appropriate measures for the prevention or minimization of hardship, loss and damage, and for relief where necessary. Unfortunately, again, while there has been much talk of 'flood management' and 'disaster-preparedness', very little has in fact been done.

---

30 It has been pointed out that the 1988 floods caused greater damage than the 1978 floods (Rs 4,630 crore as against Rs 1,455 crore; even at constant 1981–82 prices, the figure for 1988 is said to be one-third higher than that for 1978. Central assistance for flood/cyclone relief is also reported to have risen from Rs 838.3 crore in the 6th Plan period (1980–85) to Rs 2816.7 crore in four years of the 8th Plan period (1992–96). (R. Rangachari in an article in *Seminar*, June 1999.)

Irrigation

The benefits of irrigation are evident, but as a water-user it has much to answer for. As it is the largest user of water (around 80 per cent), it needs to be very efficient; unfortunately, it is in fact very inefficient. Canal-irrigation efficiency in India (around 35 to 40 per cent)[31] is well below international standards. It is true that what is lost from canals through seepage is partly recovered as groundwater recharge and as 'return flows' further down, but that is not a reason for inefficient conveyance. In any case, it is the actual application of water on the ground in irrigation that contributes more to recharge and return flows than seepage from canals. That again is not a justification for the excessive use of water in irrigation. A reduction of water-use in agriculture, and a conscious pursuit of the objective of maximum value per unit of water (more crop per drop) have to be major elements in any future water planning. If there could be even a 10 per cent saving in agricultural use, a substantial quantity of water will be released for other uses.

Secondly, injudicious canal-irrigation without regard to soil conditions, over-application of water because it is virtually free, the failure to take the groundwater table into account and inadequate attention to drainage have led to the emergence of conditions of waterlogging and salinity in many areas, resulting in valuable agricultural land going out of use. The reclamation of land thus lost is not always possible, and where feasible, it often requires large investments. A 1991 Report of a Working Group of the Ministry of Water Resources estimated the extent of waterlogged land in the country at 2.46 million hectares (MHA), and that of salt-affected land at 3.30 MHA.

Thirdly, on an average the yields of irrigated agriculture in India have been relatively low compared to what has been achieved in other countries, or even in some parts of this country; and there has been inadequate attention to increasing productivity in rainfed areas. Even the NCIWRDP's projections for the future seem fairly modest (Table 22.1).

Higher yields, which are surely achievable, will mean a reduction in the demand for water.

---

31 National Commission for Integrated Water Resource Development Plan.

Table 22.1

Average Yield	Tonnes per Hectare	
	Years	
	2010	2050
Irrigated Foodcrop	3	4
Unirrigated	1.1	1.5

Fourthly, canal-irrigation in India has been marked by a number of inequities. As waters begin to rise in the reservoir, and canal systems for taking them to the tail-end are not yet ready, the head-reach farmers have plenty of water available and tend to plant water-intensive crops. This establishes a pattern of water-use that cannot easily be changed at a later stage: by the time the full canal system is ready, much of the water stands pre-empted in the head-reach areas and there is little left for conveyance to the tail-end. This is a familiar problem in most project commands. Further, with the increasing affluence of the large farmers, their money-power begins to transform itself into political power with a potential for influencing policy-formulation and the planning, designing and location of major projects, as also their operation.

'Water Resource Development' Projects ('Major/Medium')

Some of these projects represent remarkable engineering and construction achievements (in some cases under difficult conditions), and much of the heavy equipment needed has been manufactured within India. Unfortunately, there has also been a history of poor planning and implementation, and grave failures on the human, social and environmental fronts. A study of the Indian experience with large-dam projects (for which the 'major/medium' category is a rough proxy) for the World Commission on Dams by a team (of which the author was a member) brings this out clearly. This has already been gone into in the chapters on the large-dam controversy and the framework of laws, policies, institutions and procedures within which such projects are implemented (Chapters 11 and 13 of section III). A few points not covered there in detail may, however, be added.

(i) The cost of creating irrigation potential through such projects has been steadily increasing: from Rs 1,200/hectare (ha) in the first

Plan (1951–56) to Rs 66,570/ha in 1990–92 at current prices; and from Rs 8,620/ha to Rs 29,587/ha at constant 1980–81 prices.[32] (The figures today must be much higher.)

(ii) There is a persistent gap between the irrigation potential[33] created at such cost and the extent of its utilization (Table 22.2)[34].

Table 22.2

At the end of 1995–96 (in MHA)

	Ultimate	*Created*	*Utilized*	*Gap*	*Actually irrigated (land-use statistics)*
Maj/Med	58.46	32.20	27.45	4.75	
Minor (surface)	17.38	12.10	10.72	1.38	
Minor (grw)	64.05	44.42	40.83	3.59	
Total	139.89	88.72	79.00	9.72	70.64

(iii) Resource constraints, an unsound plan/non-plan distinction and an inbuilt preference for new construction over the efficient running of what has been built, have together resulted in under-provisioning and neglect of maintenance. Systems built at great cost fall into disrepair, and there is a failure to provide the planned service.

Conflicts

As seen in the Ravi–Beas and Cauvery dispute cases, inter-State disputes over river waters are becoming intractable, and the constitutional conflict-resolution mechanisms do not seem to be working well. Conflicts could also arise between uses, between users, between areas, between classes: these are not acute yet, but could become so. There is also the possibility of conflicts between the people and the state, as has

32 Ninth Plan Working Group on Major/ Medium Irrigation Sector.
33 'Irrigation potential' is a problematic concept, but nevertheless the 'gap' between created and utilized potential cannot be dismissed as unreal.
34 National Commission for Integrated Water Resource Development Plan.

happened in the Narmada (Sardar Sarovar Project) and Tehri cases. Further, community initiatives are hampered by the 'eminent domain' claimed by the state. All this has been gone into elsewhere in this book.

Groundwater

An overview of problems relating to groundwater was given in the chapter on 'Groundwater Legislation' (Chapter 9, section II) but for convenient reference the following brief account is provided here, at the cost of some repetition.

(i) There has been over-extraction (mining) of groundwater leading to depletion in some areas, and salinity ingress in coastal zones (e.g., in Gujarat). On the other hand, there is a situation of rising water tables and the emergence of waterlogging and salinity in other areas (e.g., in the Sharda Sahayak command in Uttar Pradesh).

(ii) Water markets tend to emerge in the context of groundwater extraction through tubewells and borewells, and they serve some useful purposes, but there are dangers of unsustainable extraction as also of inequitable relationships between sellers and buyers.

(iii) The answer to both (i) and (ii) may be claimed to lie in regulation, but this has so far not been found feasible because of political factors and the legal problem of easement rights. Under the directions of the Supreme Court the Central Groundwater Authority has been established, but it is not yet clear how it will evolve and operate, what kind of regulation it will attempt, and with what success.

(iv) There are problems of pollution/contamination of aquifers (fluoride, arsenic).

(v) There is a hypothesis that there are deep aquifers under artesian conditions in the Gangetic plains, but this remains uninvestigated.

Water Quality

There are pollution control laws and institutions, but these have not been able to prevent the growing pollution and contamination of water sources and systems, which in effect makes much of the 'available' water resources unusable. This is in fact as great a threat (if not greater) to security as the 'scarcity' about which alarm bells have been ringing. What needs to be done is clear enough (prescription and continuous review of standards; their enforcement, not forgetting the cumulative impact of individual clearances and permits; making the polluter pay; adopting and moving towards clear, time-bound goals in regard to desired water quality, and so on), but not much of this has begun to happen as yet.

(Incidentally, pollution and contamination are not respecters of political boundaries. The countries of the region have to agree on common standards and on trans-boundary water quality protocols. Conflicts have arisen in the past over water-sharing; but water quality may well become the focus of even sharper conflicts in the future unless clear inter-country understandings are reached and appropriate institutional mechanisms are provided for ensuring compliance with such understandings.)

Waste of Water

There is a waste of water in every use: agricultural, industrial, municipal, domestic. There is a complete absence of a sense of scarcity, and this is aided by a gross under-charging of water, whether for irrigation or in urban water supply. On the subject of irrigation charges, the Report of the Committee on the Pricing of Irrigation Water (the Vaidyanathan Committee) set up by the Planning Commission was submitted in 1992, but its recommendations still remain unimplemented. The National Commission for Integrated Water Resources Development Plan, in its Report of September 1999, has reiterated those recommendations with some modifications and additions. However, it cannot be said that there has been a significant change in the prevailing attitudes to water, whether in the matter of pricing or in that of conservation and economy.

23

Objectives for the Future

Having regard to the summary of problems and weaknesses given in the last chapter, what needs to be done? Policies and plans for the future must be guided by a vision of the kind of world that we would like to see. A detailed vision statement is not attempted here, but there will be general agreement that it should be a vision of a sane (i.e., not necessarily affluent, but prosperous in a moderate sense), humane, caring, sensible, equitable, just, efficient, harmonious and sustainable world. Keeping that in mind, our objectives for the future (relating to water) can be enumerated as follows:

- Ensure access to safe drinking water for all.
- Ensure adequate availability of water for agriculture, industry and urban centres (with due regard for efficiency, economy and equity).
- Find appropriate answers for drought-prone areas and arid zones.
- Foster consciousness of scarcity, promote conservation and minimize waste.
- Improve and maintain water quality, control pollution and protect water sources.
- Protect and preserve natural environment/ecological system; preserve integrity of rivers and maintain river regime.
- Ensure equity—between groups, between generations and between species.
- In particular, reduce burden on women and give them a voice in water planning and management.
- Minimize conflicts and hardships and provide means of resolution/redress.
- Help people to cope with floods and minimize damage.

How will those objectives translate into policies and action plans? With reference to most of the matters discussed, the diagnoses themselves indicate what needs to be done. On two topics, this book will not enter into a detailed discussion, but will merely draw the reader's attention to useful material available elsewhere.

(i) On flood-related issues, there are excellent articles by R. Rangachari in a special issue of the journal *Seminar* and by D.K. Mishra in the *Economic and Political Weekly*.
(ii) On the issue of water quality, there is a useful chapter in the book *Cooperation on the Eastern Himalayan Rivers*.

Details of the above will be found in the select references and bibliography at the end of this book. The author does not have much to add to those writings.

In the ensuing chapters, we shall review assessments of future water requirements and reflect on the dilemmas of 'water resource development'.

24

Assessments of Future Water Requirements

Introductory

In recent years there have been several studies of India's water require-
ments in the future years. The present review covers three important
ones: (i) the estimates made by the Working Group on Perspectives of
Water Requirements (WG) of the National Commission for Integrated
Water Resources Development Plan (NCIWRDP) and adopted by the
Commission in its Report to the Government of India (September
1999); (ii) 'India Water Vision 2025' prepared by the India Water
Partnership, an informal body loosely affiliated to the Global Water
Partnership, as part of the preparations for the World Water Forum held
at The Hague in March 2000; and (iii) the final report (October 2000)
of a study of sustainable water resources development by Kanchan
Chopra and Biswanath Goldar of the Institute of Economic Growth,
Delhi.[35] (Certain other estimates such as those by the Ministry of Water
Resources or the Indian Water Resources Society have been mostly taken
into account in the above studies and need not be separately reviewed.)

Water Availability

Before reviewing the projections of requirements, it may be useful
to remind ourselves of the picture of availability as given by the
NCIWRDP. The average annual precipitation by way of rain and snow
over India's landmass is 4,000 billion cubic metres (BCM),[36] but the

---

35 Full references to all these will be found in the bibliography.
36 Billion cubic metres is the same as km^3 or cubic kilometrs.

annual water resources of the country, measured in terms of the flows at the terminal points in the river systems, have been estimated by the NCIWRDP as 1,953 BCM. The groundwater availability has been put at 432 BCM.[37] As against those 'availability' figures, the NCIWRDP estimates that the annual 'usable' water resources of the country are 690 BCM of surface water and 396 BCM of groundwater, making a total of 1,086 BCM. The present quantum of use is put at around 600 BCM.

Working Group of the NCIWRDP

In its Report, the WG has made estimates of water requirements for the years 2010, 2025 and 2050. Ignoring certain preliminary chapters (introduction, methodology, overview of water resources and conceptual aspects), let us proceed straight to the estimation of water requirements. The requirements for different purposes (irrigation, drinking water, municipal use, industrial demand, etc.,) are estimated and then these are aggregated to arrive at the total requirements.[38]

Irrigation

The requirements of water for irrigation are derived from projections of food requirements. The need for self-sufficiency in food is a basic assumption in the study. The Report observes: 'The overriding consideration in estimation of the water requirement for irrigation is self-sufficiency in food production at the national level. The WG also maintains that the present ratio of food and non-food production in terms of area sown is sustainable on physical and socio-economic considerations.' As requirements of food production would depend mainly on the country's population, per capita income and changes in dietary habits, the Report proceeds to deal with these matters.

The WG refers to population projections by (i) K.S. Natarajan, Annamalai University, December 1993; (ii) 'World Population Prospects—the 1994 Revision', United Nations, New York, 1995;

---

37 Treating those two figures as distinct could conceivably involve some double counting in view of the complex interactions between surface water and ground-water.

38 This is referred to as the 'building block approach'—a rather impressive name for a simple proposition.

(iii) 'Population Projections of India and States 1996–2016', Registrar General India, Census of India, Government of India, Ministry of Home Affairs, New Delhi, 1996; (iv) 'Prospective Population Growth and Policy Options for India—1992–2001' by Visaria L. & Visaria P., The Population Council, New York, 1996. After a brief account of these projections, the WG adopts 'population projections of UN (low variant) and projection made by Visaria & Visaria as low and high population growth scenarios for estimation of food demand and water requirements for drinking purpose.' The figures are given below in Table 24.1.[39]

Table 24.1

(in millions)

	Year			
Source	*2000*	*2010*	*2025*	*2050*
Low (UN 1994, Low Variant)	1,013.5	1,156.6	1,286.3	1,345.9
High (Visaria and Visaria 1996, Standard)	995.0	1,146.0	1,333.0	1,581.0

For the rural-urban distribution, the WG again uses a combination of UN, Registrar General India and Visaria & Visaria figures and comes up with the following projections of urban population (Table 24.2).

Table 24.2

	Year		
	2010	*2025*	*2050*
Low	367	476	646
High	387	603	971

Based on the above, and taking into account all the relevant consid-erations (rural and urban per capita consumption patterns, dietary changes, expenditure distribution, socio-economic factors, etc.,) the WG then goes into the demand for food, drawing upon projections by G.S. Bhalla and Peter Hazell and a study by C. Ravi commissioned by

---

39 This table and others in this chapter are drawn from those in the reports under review, but are not necessarily full reproductions.

the WG itself, and adding provisions for livestock-feed requirements, storage and transportation losses, seed requirements, and carry-over for years of monsoon failure, it puts the total foodgrain requirements for the country in the year 2050 at 494 million tonnes (high) and 420 million tonnes (low).

From those projections the WG proceeds through considerations of land-use, cropping patterns, foodgrain yields, irrigation depth and efficiency, etc., to the following projections of the water demand for irrigation (in BCM):

Year 2010: 543 (low)/557 (high);
Year 2025: 560 (low)/611 (high);
Year 2050: 628 (low)/807 (high).

Domestic Use

After considering various norms (Gleick, WHO, etc.,) for rural and urban water supply, the factors involved, the uneven distribution of actual supplies, etc., the WG suggests the following norms (Table 24.3).

Table 24.3

Unit: litres per capita per day (LPCD)

Category	Year		
	2010	2025	2050
Class I City	220	220	220
Other than Cl. I Urban	150	165	220
Rural	55	70	150

The WG adopts the following 'scenarios' for estimating domestic water requirements (Table 24.4).

Table 24.4

Scenario	Population and Rate of Urbanization
High Water Requirement	Population as per Visaria & Visaria and high rate of urbanization
Low Water Requirement	Population as per UN low variant and low rate of urbanization

On the above basis, the national high and low water-demand projections made by the WG are given in Table 24.5.

Table 24.5

Unit: BCM

	Year		
Scenario	*2010*	*2025*	*2050*
Low Demand	42	55	90
High Demand	43	62	111

The water requirements of the bovine population are estimated at 4.8, 5.2 and 5.9 BCM for 2010, 2025 and 2050.

Industry

On the basis of assumptions regarding the overall rate of industrial growth, the rates of growth of different industries, the per-unit water consumption rates, etc., and emerging water-saving technologies, the WG estimates the water requirements of the industrial sector as 37, 67 and 81 BCM in the years 2010, 2025 and 2050.

Power

With reference to projected installed power-generation capacities, standards of water use, etc., and with a provision for the requirements of nuclear, wind/solar and gas-based plants, and taking into account emerging technologies, the WG estimates the water requirements of the energy/power sector, for the low and high demand 'scenarios', as 18 and 19 BCM; 31 BCM and 33 BCM; and 63 BCM and 70 BCM for the years 2010, 2025 and 2050. (Hydropower stations in general are regarded as not having any consumptive use of water, except for some releases of unused water into the sea by coastal plants.)

Inland Navigation

The Report surveys the present situation in regard to inland navigation and the prospects of development, takes into account the norms recommended by the National Transport Policy Committee (1980), and proceeds to consider the water requirements. These are 'mostly expected to be met by seasonal flows in various river systems or canals', but where a river is dammed, 'some water would be required to be released ... for keeping the waterways navigable'; but 'it is also likely that minimum flows released from essential environmental consideration may exceed the flow needed for navigation'. In the light of the above, the WG provides 7, 10 and 15 BCM of water for inland navigation in the years 2010, 2025 and 2050 respectively.

Flood-Moderation

The WG comes to the unsurprising conclusion that flood-moderation does not need water!

Environment and Ecology

Under this head the WG considers mainly two requirements: for (i) afforestation and tree-planting, and (ii) pollution control and main-tenance of water quality. The first is expected to be taken care of from precipitation and soil moisture, and so the WG makes no separate provision.[40] The second runs into difficulties of estimation. The WG therefore makes 'token' provisions of 5, 10 and 20 BCM for the years 2010, 2025 and 2050.

Evaporation Losses from Reservoirs

These are estimated at 36, 42 and 65 BCM for the years 2010, 2025 and 2050.[41]

---

40 Implicit in this is the assumption that the 'available' water is the quantum that is stored or diverted through dams and barrages, and that use direct from precipitation or soil moisture need not be counted against that availability. If we define 'available' or 'usable' water to include local augmentation through *in situ* conservation, rainwater harvesting and watershed development, then the use of that water will have to be reckoned against that availability.
41 These numbers include evaporation from minor projects taken at 25 per cent of live storage capacity as against 15 per cent in the case of large projects. The basis for this is not clear.

Total Water Requirement

The country's total water requirements are estimated as given in Table 24.6 (figures in BCM):[42]

Table 24.6

	Year		
	2010	2025	2050
Low	694	784	973
High	710	850	1,180

(The WG has also made basinwise and Statewise estimates as first approximations on the basis of 'heroic' assumptions. In fact it says that all the estimations are approximate and need frequent reviews.)

Emerging Issues

In its final chapter, the Report highlights some important issues:

- the probability of a decline in per capita water availability and of the emergence of water-stress conditions;
- huge disparities in availability among the regions and States;
- shift in the requirements between rural and urban areas;
- increasing pollution (which may become massive) of both surface water and groundwater;
- the likelihood of serious conflicts between different uses of water, different areas, different States;
- inequities in water availability and in payments for water, with the poor often paying more than the rich; and
- institutional weaknesses and absence of beneficiary participation.

The WG then proceeds to make a number of recommendations:

- integrated water resource management;
- prevention of pollution of water resources;

---

42 These figures are from Chapter 14. The figures in the executive summary of the Report are somewhat different. It is difficult to explain this.

- economy in use and resource conservation;
- promotion of rainwater harvesting;
- the recycling of used water;
- better water management in irrigation (this is discussed at some length);
- rationalizing irrigation water rates;
- organizational reforms such as beneficiary participation; and
- other measures, including the use of saline or brackish water, conjunctive use of surface water and groundwater and so on.

National Commission

The NCIWRDP in its Report of September 1999 adopts and incorporates the WG's Report in a condensed form. The numbers are the same, except that in the case of industrial demand for water the Commission gives two figures for the year 2050: 81 BCM and 103 BCM. The latter figure is on the basis of present rates of use, whereas the former assumes the adoption of water-saving technologies.

In summing up the position of water availability and requirements, the NCIWRDP repeats the WG's point about the approximate nature of the estimates and the need for frequent reviews. While not taking an alarmist view of the 'fragile balance between the aggregate availability and aggregate requirement', it emphasizes the importance of ensuring that the 'low' projections prevail. It stresses the need for 'utmost efficiency' in water-use and for equity in access as between regions and between sections of the population. (There is a reference to the need for water transfers from surplus areas to those of scarcity, though the NCIWRDP does not make any strong recommendations about major 'inter-basin transfers'.)

India Water Vision 2025

Here again, we may skip the preliminaries such as the background, an account of the 'vision' processes, vision elements, vision 'drivers', the argument about 'food security *versus* free trade', etc., and go straight to the projections. (This Report, hereafter 'IWV', also comes to the conclusion that food security is very important.)

'Scenarios'

The IWV makes projections under three 'scenarios': 'BAU' ('business as usual', meaning that we carry on as before),[43] 'Improved Management' (meaning that some incremental changes and improvements take place) and 'Sustainable Water World' (implying somewhat larger changes in the interest of 'sustainability').

Agriculture

The IWV, like the WG and the NCIWRDP, proceeds from population and urbanization projections, food requirements, etc.—in fact it draws upon the work of the NCIWRDP—and comes to the conclusion that as the Net Sown Area is unlikely to increase significantly, and as the area under foodgrains may at best remain at the present level, increases in food production have to come from stepped-up productivity; and that (in the absence of any dramatic technological developments) this would require an extension of irrigation. The IWV observes that 'depending upon the level of demand assumed, between 33 and 48 million hectares of foodgrain area currently under rainfed conditions would need to be brought under irrigated conditions'. Using the same 'delta' figures for irrigation water as the NCIWRDP, the IWV estimates the irrigation water requirement for food production in the year 2025 as 507, 558 and 600 BCM for the three 'scenarios' respectively. After a brief discussion of 'non-food production', IWV puts the total water requirement for irrigation in the year 2025 at 730.6, 805.6 and 866.6 BCM under the three scenarios respectively. This would represent 67 per cent, 74 per cent and 80 per cent respectively of the country's total available (usable) water resources of 1,086 BCM.

The Report then examines the magnitude of investment needed for meeting the projected irrigation water-demand, and what needs to be done to ensure the financial viability and sustainability of irrigation investments. We need not go into these matters here.

---

43 I do not know who first used the inelegant expression 'business as usual' to mean 'carrying on as before' or 'making no significant departures from past policies and practices', but it has gained currency. I am obliged to use it though I would have preferred not to.

There is then a section devoted to 'livelihood security' which deals with the contribution that the extension of irrigation will make to the generation of employment. It also refers to such subjects as marketed surplus of foodgrains, foodgrain prices, impact of irrigation on poverty and on income distribution, the development of aquaculture and so on: we seem to have moved quite far from the subject of water. This discussion does not translate into water requirements, these having been already covered under the head of irrigation.

Domestic and Municipal Use

The next section deals with 'water for health security' but while the importance of water in the context of water-borne, water-washed, water-based and water-related diseases is explained and the need for safe water and sanitation is stressed, this discussion again does not directly lead to any figures of water requirements. In fact this section is a preliminary to a discussion of water supply and sanitation, but that discussion also does not lead to projections of water demand. Water requirements for 'domestic use'—that general description subsumes the needs dealt with in earlier paragraphs—are indeed gone into, but independently of the earlier paragraphs; they are derived from certain per capita norms and certain assumptions regarding population growth and the rate of urbanization. Paragraphs 3.3.1 ('diseases associated with water'), 3.3.2 ('need for safe water and sanitation') and 3.3.3 ('water supply and sanitation in India') contain discussions that are not integrally related to the consideration of water requirements in paragraph 3.3.5.

(Incidentally, it would appear that someone at some stage cut up some pages of the text of the Report with a pair of scissors, shuffled the pieces, and then put them together in random fashion. How else can one explain the fact that in the middle of the discussion of water supply and sanitation there is an unrelated paragraph on irrigation infrastructure (p. 49) which ought to have come earlier, or that between paragraph 3.3.3, which deals with water supply and sanitation, and paragraph 3.3.5, which deals with water for domestic use, there is a paragraph (3.3.4) that deals with hydropower?)

Returning to water supply for urban and rural areas, IWV makes a reference to the NCIWRDP's norms and projections for the years 2025 and 2050, but there is some difficulty in relating these to the numbers and tables in the Commission's Report. We may leave that aside. IWV finds that its own norms in the 'Sustainable Water World' scenario are in conformity with the 'high' per capita norms of the NCIWRDP. It suggests the adoption of the Commission's low population projection. On that basis it estimates the water requirements for domestic use in 2025 at 70 to 78 BCM under the 'Sustainable' scenario, and adds that as this is only 10 per cent of the total requirements for all uses, and as the National Water Policy 1987 accords the highest priority to drinking water, these requirements are likely to be met. (Under the BAU scenario, IWV projects lower requirements: 30 to 50 BCM.) This is followed by a discussion of investment requirements which we need not go into.

Ecological Security

The next section deals with ecological security. It briefly discusses the degradation of the environment; the environmental impacts of water resource development projects; the drying up of rivers and the need to maintain minimum flows, and the problems of water pollution and the maintenance of water quality. It then projects (in qualitative terms) the 'BAU' and 'sustainable' futures. Turning to water requirements, it quotes—without disagreement—the NCIWRDP's projections of 67 BCM for afforestation and 10 BCM for maintaining minimum flows in rivers in the year 2025.

Total Water Requirements

In the light of all these discussions, IWV arrives at a total water require-ment of 1,027 BCM (gross) for the year 2025 as follows: irrigation 730; water supply 70; ecological 77; industry 12; energy 30. (There is some-thing missing here, as these numbers add up only to 919 BCM!) Es-timates of the investment needs are then given. A discussion of policy instruments and institutional needs follows.

Chopra and Goldar

This is the most careful, comprehensive and rigorous of all the studies reviewed here. Its richness, complexities and density make it difficult to summarize in any meaningful sense. What follows is merely a very selective presentation of some salient and relevant details.

'Scenarios'

Passing over the conceptual discussion of sustainability and of the indicators and dimensions of sustainability in relation to water resources, let us proceed straight to the three scenarios adopted by the authors, namely, Business As Usual (BAU), High Growth (HG) and Sustainable Scenario (SS). The names are self-explanatory. The BAU and HG scenarios do not assume any improvements in water-use efficiency, whereas the SS scenario does. (The BAU scenario includes a BAUST variant which gives Statewise projections, which will not be gone into here.)

Water Availability

The study makes projections for the year 2020. In so far as the availability of water is concerned, the study accepts the CWC/NCIWRDP figures for surface water: available water resources 1,953 BCM, and usable water resources 690 BCM. The groundwater availability is defined as 'replenishable groundwater' and put at 420 BCM. (It is not clear whether this corresponds to the NCIWRDP's 'potential' figure of 432 BCM or its 'usable' figure of 396 BCM.) The water available for irrigation is estimated as 315.98 BCM of surface water and 342 BCM of groundwater. (Land degradation is expected to affect this availability by 6 to 7 per cent, unless 'sustainability' measures are taken.)

Agricultural Requirements

Two population estimates for the year 2020 are mentioned: the revised UN estimate of 1,272.2 million and that of the Technical Group of the Population Foundation which is 1,345–1,355 million. The study adopts the revised UN estimate. The urban component of this is taken as 40 per cent. The water requirement for agriculture is arrived at *via* projections of food demand and of demand for vegetables and fruit, assumptions/estimates in regard to rainfed area, irrigated area requirements, productivity on irrigated and rainfed land, etc. The kinds of interventions needed for sustainable development and the investments that are required are gone into. The picture that emerges in regard to the water requirements for agriculture in 2020 under the three scenarios is given in Table 24.7.

Table 24.7

	UNITS	BAU	HG	SS
Agricultural Growth Rate (food and non-food)	Growth Rate	4.69	4.98	4.98
Irrigated Land	Million hectares (MHA)	112	122	122
Water Requirements: Surface	BCM	677.3 248.96	804.2 804.2 (s+g combined)	768.37 626.37 (s+g combined)
Groundwater		428.34		
Additional Run-off Captured[44]		0	0	142
Water Deficit	Percentage	2	22	0

The Report observes: 'The HG scenario in agriculture is feasible in terms of its water requirements only if sustainability investments are undertaken and result in better water-use efficiency, capturing of more run-off to augment water availability and prevent quality deterioration.'

Household Requirements

On the basis of a projected population of 1,272.2 million by 2020 (revised UN); an urban component of 40 per cent, with 80 per cent of that component in Class I cities (Asian Institute of Transport Development); WG and NCIWRDP supply norms with a moderation of the rural area norm to 70 LPCD; and an upward adjustment of the NCIWRDP estimate of the requirements of livestock (taken as a part of the rural water requirement), the Report puts the total requirement for households (which was 35.2 BCM in 1995) at 67.5 BCM by 2020. The Statewise distribution of this is also given.

---

44 Through rainwater harvesting and watershed development. See footnote 40.

Power-Generation

On the basis of an estimated installed capacity of 509,136 MW in 2020, and using the norms of the WG for water consumption for power-generation (taking hydropower as not having any consumptive use of water), the Report expects the requirement of water for power to increase from 1.51 BCM in 1995 to 8.19 BCM in 2020. This is fairly close to the NCIWRDP's estimate of 10 BCM for 2025.

Industry

The total water requirement of manufacturing industry is projected to increase from 7.8 BCM in 1995 to 27.91 BCM (BAU), or 41.58 BCM (HG). The Report assumes that the water consumption norms will remain the same as they were in 1995.

Total Requirements of Non-Agricultural Sectors for 2020

The Report presents the following 'scenarios' for the non-agricultural sectors for 2020 (Table 24.8).

Table 24.8

	BCM		
Requirements	BAU	HG	SS
Households	67.52	67.52	45.01
Power	8.19	11.47	5
Industry	27.91	41.58	27.72
Total	103.62	120.57	77.73
Evaporation Loss	42	42	42
Environmental/	78	78	78
Ecological Requirement			
Total Requirements	223.62	240.57	197.73

These are within the availability for non-agricultural sectors as estimated by the Report (374.34 BCM). At the State level, however, the balance is expected to be not very comfortable in a number of States.

Total Water Requirements

There are two sections entitled 'Water Pollution' and 'Interventions for Sustainable Water-Use in Non-Agricultural Sectors' which we need not go into here. We proceed straight to the concluding section. Taking both agricultural and non-agricultural water requirements, the total water requirements in 2020 are given as 920.92 BCM (BAU), 1,004.77 BCM (HG) and 964.09 BCM (SS). (The arithmetic actually seems to work out to 900.92, 1,004.77 and 966.1 BCM.) These are within the overall availability (surface and groundwater together) of 1,110.566 BCM projected for 2020. (The NCIWRDP's figure is 1,086 BCM.) The position therefore seems manageable, but this assumes that interventions on the demand management and supply augmentation sides will in fact be undertaken and will be successful. The water-saving will be the difference between the BAU and SS projections for the different uses, and the augmentation will be the projected additional run-off capture of 142 BCM.

That is the overall national picture. There will be variations among the States. Some States (e.g., Gujarat, Maharashtra, Haryana, Tamil Nadu, Andhra Pradesh) are expected to experience acute shortages in groundwater or surface water or both.

The concluding paragraph reiterates the need for two kinds of interventions: those for improving water-use efficiency in agriculture, and those for improving quality in relation to the household and industrial sectors.[45] Some indications of the investments needed for such investments are given. Institutional changes are also stressed. The Report concludes with the observation that without such steps the HG scenario will not be compatible with sustainability.

Some Comments

The studies reviewed here differ in some ways but are not widely divergent. Broadly speaking, all three agree on the quantum of water availability (with a minor difference in regard to groundwater between the WG and Chopra/Goldar and a significant difference in relation to

---

45 Surely water-use efficiency needs to be improved in the household and industrial sectors as well.

additional run-off capture, which we shall revert to shortly); their population projections are mainly based on the revised UN estimates (1994), and they adopt roughly similar percentages for the urban component; they envisage an addition (of different degrees) to the irrigated area; the norms that the NCIWRDP adopts for rural and urban water supply are used by IWV and Chopra/Goldar with some modifications; there are some minor differences in other respects (e.g., the provision for environment/ecology, etc.,). The total water requirements projected by the WG and NCIWRDP (973 to 1,180 BCM in 2050) are relatively lower than IWV's projections (1,027 BCM in 2025) or Chopra/Goldar (920.92 BAU, 1,004.72 HG, and 964.9 SS in 2020).

All three studies envisage an effort at 'sustainability'. While IWV and Chopra/Goldar project a separate 'sustainable' scenario (on the basis of certain measures and investments), WG's and NCIWRDP's projections assume that certain steps to ensure economy, efficiency and conservation will be taken, and predict a fragile balance between supply and demand on that basis. IWV and Chopra/Goldar also seem to adopt a similar position of a cautious but not an alarmist view of the future.

Is such a position warranted? There are two divergent views on this question. One is that the NCIWRDP has been somewhat complacent and failed to sound the alarm bell. One has heard Dr Y.K. Alagh say this at several meetings. In fact he quotes Dr S.R. Hashim (the Chairman of the NCIWRDP from 1997 to 1999) himself as sounding a note of caution (not in the Report, but in a lecture). The other view, questioning the prediction of a crisis, is that of the Centre for Science and Environment (CSE): they argue that if local, community-based water-harvesting is undertaken extensively all over the country (wherever it is feasible) there will be no crisis. They circulated a paper to this effect at the Hague Forum.

It is difficult to quantify precisely the hydrological consequences of local water-harvesting on an extended scale. The Chopra/Goldar study projects an 'additional run-off capture' of 142 BCM through these means, but the basis for this estimation has not been given. Assuming that 142 BCM of run-off is captured through local water-harvesting and watershed development, will this improve or reduce the availability lower down? It is possible that water levels in wells and aquifers will rise: this has been observed in various places. But will the flows in the streams and rivers be reduced? Rajendra Singh keeps pointing out that the flows in the Arvari (in Alwar district in Rajasthan) have in fact increased, but is that finding generalizable? To the extent that the run-off is intercepted at the earliest stages, is it not possible that the run-off that drains into the river will stand reduced? Some of the captured water will no doubt become 'return flows', but that will be necessarily a percentage of the water intercepted. Some of it will also get polluted or contaminated in the processes of use. If we take all the pluses and minuses together, can we say that the net result will be a substantial addition to 'available' and 'usable' water (in the NCIWRDP's sense)? As the 'usable' surface water (690 BCM) is only a fraction of the 'available' water (1,953 BCM), and as that in turn is only a fraction of the precipitation (4,000 BCM), it seems reasonable to assume that the capture of more rainwater will add to the availability of water for use. However, one is not aware of any clear answer to this question, other than the figure of 142 BCM mentioned in the Chopra/Goldar study.

Leaving that aside, what needs to be noted is that despite the differences between the three studies, there are also commonalities. All of them operate within the framework of a demand-supply calculus; they share similar ideas of 'growth' and 'development', and they have similar understandings of 'sustainability' (some measures of efficiency and economy on the demand side and a degree of augmentation on the supply side). None of them envisages radical departures from the past. The 'sustainable' scenario that they talk about is largely the BAU scenario with some efficiency/conservation measures added. Essentially, the approach is to proceed from projections of demand to supply-side answers. Environmental and ecological concerns are seen not as limits on our draft on nature, but as yet another category of 'demand' for which an 'allocation' has to be made, taking the total water requirements higher. One would have expected the water requirement to be lower in the 'sustainable' scenario than under BAU, but it is the other way round in the Chopra/Goldar study, which seems rather strange. Moreover, that study blurs the distinction between the HG and SS scenarios when it suggests that the former is feasible if sustainability investments are undertaken (*vide* quotation cited earlier). One would have thought that 'HG' would be *ipso facto* unsustainable, and that SS would be significantly different—different in kind—from HG.

The view of this writer is that if we go by the demand-supply calculus and by the prevailing notions of 'development', a crisis is inevitable. It can perhaps be averted, but not merely by supply-side solutions, whether large-scale and centralized or small-scale and local: radical changes in our ways of living are called for. This is not seriously discussed in the

WG/NCIWRDP/IWV Reports (though some *pro forma* references are made to 'lifestyle changes'). Even the Chopra/Goldar study, which is very good indeed (as mentioned earlier), begins with economic growth (as commonly understood) at certain desiderated rates, refers to certain rates of water-use and norms of consumption, projects water demands on that basis, tries then to bring in the 'sustainability' criterion and applies corrections that range from limited to modest to significant.

That comment is likely to be countered with the question: 'Is that not the right thing to do? How else would you have proceeded?' One would have liked to envision a future on lines very different from the prevailing notions of 'economic growth' and 'development'. (One is not advocating Gandhian austerity, but a goal of modest prosperity.)

A radical departure from current thinking will undoubtedly be fraught with serious difficulty and will pose formidable challenges. We could prefer not to face them; we could assume ('realistically', as some may say) that no more than moderate improvements in efficiency and economy in water-use are likely to happen, and we could assume further (even more 'realistically') that there will be no change in our ideas of 'development'; under those assumptions there will indeed be a horrendous water scarcity necessitating massive supply side projects. We must then also 'realistically' acknowledge that sustainable development is an impossibility, and that planet Earth (and along with it humanity) is doomed. Instead of perplexing ourselves with such apocalyptic forecasts, will it not in fact be *more realistic* to recognize that given the precious nature of this life-sustaining element and its finite supply on this planet, a tremendous effort needs to be made not merely at efficiency, economy and conservation,[46] but also at a reordering of our lives? Alas, there *are* no easy answers to that conundrum. We shall return to this theme (even at the cost of some repetition) in Chapter 25.

---

46 The reference here is to water for irrigation, industry, etc., and not to water for drinking, cooking and washing, though even here there is much scope for economy in the case of the middle and upper classes. This is argued in Chapter 25.

25

The Dilemmas of 'Water Resource Development'

Introductory

The term 'water resource development' or 'WRD' has acquired a special meaning among those concerned with water resources in this country and perhaps elsewhere (hence the inverted commas in the title). In our current usage, WRD generally means large supply-side storage/diversion projects. (It also means groundwater extraction and use, but that subject has been discussed elsewhere. This chapter will be concerned mainly with surface water and will make only a brief reference to groundwater.) As mentioned earlier, the standard engineering response to temporal and spatial variations in the availability of water is to propose: (*a*) the storing of river waters in reservoirs behind large dams to transfer water from the season of abundance to that of scarcity, as well as from good years to bad, and (*b*) long-distance water transfers from 'surplus' areas to water-short areas. Against that background, the water resource planners in this country (as in many other countries) feel that a number of large dam-and-reservoir projects are called for and should be taken up urgently. In its response to the Report of the World Commission on Dams the Government of India, Ministry of Water Resources, observed as follows:[47] 'Having made impressive strides since Independence in developing our water resources, India proposes to continue with its programme of dam construction to create another 200 billion cubic metres (BCM)

---

47 Government of India, Ministry of Water Resources, Letter No. 2/WCD/2001/DT(PR)-Vol. III dated 1 February 2001 to the Secretary General of the World Commission on Dams.

of storage in the next 25 years or so to ensure continued self-sufficiency in foodgrain production and to meet the energy and drinking water needs of a growing population.'

Unfortunately, we run into several difficulties here. It might have been noticed that the title of the chapter refers to 'dilemmas' in the plural. There are four main dilemmas (with connections among them): the money dilemma, the markets dilemma, the environmental dilemma and the human rights dilemma.

The Money Dilemma

Let us consider the money dilemma first. The India Water Vision 2025 exercise in the late 1990s (under the auspices of the Global Water Partnership) for presentation at the Hague Forum of March 2000 postulated an investment of Rs 5,000 billion in 25 years, or Rs 200 billion per year.[48] The National Commission for Integrated Water Resources Development Plan's (NCIWRDP) rough estimates of amounts needed for completing spillover projects are Rs 70,000 crore in the Tenth Plan and Rs 1,10,000 crore in the Eleventh Plan.[49] Not only does this leave no scope for new major projects, but the difficulty of finding funds of this order necessitates a severe selectivity even in regard to the continuance of what are called 'ongoing projects'. The Report devotes a whole chapter to the 'prioritization' of these.[50] Some very hard choices are called for: some projects may have to be accelerated, others restructured and drastically pruned, and yet others abandoned.

What we must accept in a clear-headed manner is that the actual availability of investment funds in the public sector is likely to be no more than a small fraction of the projections made. Nor should we delude ourselves into thinking that the answer lies in private sector investment. National private sector investment in this sector, if forthcoming, is likely to be marginal at best, and we have to be very wary

---

48 'India Water Vision 2025—Report of the Vision Development Consultation', India Water Partnership and Institute for Human Development, published by the Institute for Human Development, New Delhi, 2000, pp. 2 and 70.
49 NCIWRDP's Report, p. 242.
50 Report, Chapter 10.

of possible investments by foreign or multinational corporations. The point to be noted here is that the pursuit of WRD in the form of large projects is likely to be a frustrating effort and that much time will be lost in this process with uncertain prospects of success. We seem convinced that a massive construction programme is urgently necessary, but the financial resources needed are nowhere in sight: how then are we going to meet future needs as we see them?

The Markets Dilemma

We turn now to the 'markets dilemma'. If engineers and administrators tend to argue for supply-side projects, 'liberal' economists and officials of the multilateral financial institutions tend to argue for water markets. To them water is a commodity like any other, governed by the laws of supply and demand: if the state stops meddling and leaves it to the private sector and to the operation of market forces, then supply will meet demand, prices will be right, economy and conservation will be ensured and conflicts will be automatically resolved by the market. This route is also expected to resolve the financing dilemma that was mentioned earlier. The slogan is: 'water is an economic and social good'. This has been discussed in Chapters 7, section II (regarding perceptions), 8 (the paragraphs entitled 'economic perspectives'), 9 (the part relating to 'water markets' in the context of groundwater use), and 10 (the paragraphs relating to tradable property rights). In the light of the observations in those chapters, the dilemma here is a multiple one:

(i) in the domestic context, how to allow limited and regulated water markets to function without inequity and injustice, without danger to the resource and without transferring rights and control substantially away from the community to the state or other agencies; and how to accept (and indeed insist on) a stringent commercial approach in the context of supply for industrial use, for commercial agriculture and for luxury consumption by the affluent, without carrying the 'commodification'[51] of water too far; and

---

51 A term that has come into use, though (if a special term is considered necessary) one would prefer 'commoditization'.

(ii) in the international sphere, how to ensure that the few instances of international trade in water that now occur do not burgeon into a massive bulk trade in water as in the case of oil; and how to protect the rights of the poorer and weaker countries over their own natural resources from predatory corporate giants.

The Environmental Dilemma

Let us proceed to the environmental dilemma. As we saw earlier, the Ministry of Water Resources has put before the nation an ambitious target of construction of 200 BCM of storage capacity in the next 25 years. It is not necessary to go into the details of the numerous projects adding up to that figure. It is clear enough that even if a part of that target is attempted, the impacts on the natural environment and the ecological system will be very substantial. The environmental impacts will of course vary in range and severity from case to case, but most such projects have some common and inescapable consequences. This has been gone into in the chapter on the large-dam controversy (Chapter 11, section III). As we saw there, some of those effects cannot be remedied or even mitigated; and in some cases efforts at the mitigation of or compensation for environmental impacts may in turn create further problems. Further, it is clear from past experience that all the consequences and ramifications arising from the damming of a river cannot really be fully foreseen and planned for. As was argued further, environmental impact assessments (EIAs) and cost-benefit analyses are undependable and unsatisfactory answers to these problems. (This is not an argument against EIAs or cost-benefit analyses; they are necessary; this is merely a plea that their limitations as the basis for investment decisions be recognized. Some suggestions in this regard will be put forward later.)

The dilemma here, i.e., in the environmental context, is that we think that we can 'harness' natural resources for human benefit while limiting the harm that we do, and that the 'trade-off' is in most cases positive; but this is open to serious question. We may in fact be sitting on the edge of a branch and axing it at the junction to the tree, or perhaps cutting down the tree itself. If, in undertaking major interventions in nature to increase the availability of water for use, we do not know exactly what we are doing and put the ecological system at risk, what will happen eventually to the water? If we forget for a moment the questionable calculus of supply and demand and look at 'security' from the point of view of protecting the ecological system and planet earth, we begin to realize that by building a series of large WRD projects we may not be *ensuring* security but *endangering* it.

In this context, let us take note of three fairly common statements that are often advanced with a degree of plausibility in discussions and arguments.

(i) 'Environmental concerns are all right, but they should not be allowed to come in the way of development'.

(ii) 'Economic development and environmental concerns need not be assumed to be in mutual conflict; both are important and can be harmonized'.

(iii) 'Without human beings the very word "environment" makes no sense; human needs come first and must be given priority over concerns about flora and fauna'.

These remarks reflect profoundly wrong ways of thinking. Without entering into an elaborate discussion of the fallacies involved, the following responses are offered (taking the above statements in the reverse order):

(i) The postulation of a dichotomy between humanity on the one hand and flora and fauna on the other shows a failure to understand the ineluctable relatedness of all of nature. In John Donne's famous observation that 'every man is a piece of the continent, a part of the main', the word 'man' needs to be replaced by 'form of life'.

(ii) It would be right to say that there is no conflict between economic development and the environment *if and only if* our understanding of what constitutes 'development' undergoes a radical change. Between 'development' as at present understood and the health of the natural environment, conflict is not merely possible but inevitable.

(iii) It is foolish to imagine that environmental (or to be more precise, ecological) concerns can be ignored and 'development' somehow achieved.

At this juncture, and by way of a digression, it may not be out of place to make a reference to two questions posed by a distinguished professor at a conference on the eastern Himalayan rivers at Kathmandu some years ago. Referring to the sustainability debate, he said that the concerns could be captured in the following two questions: (1) How far can we go in our draft on natural resources before we cause harm to future generations? (2) How far can we go in dumping pollutants and contaminants in water, soil and the atmosphere before we cause irreversible damage? At first sight these questions seem well formulated, but if we consider them more carefully, we realize that they ask not a *positive* question ('how can we live in harmony with nature?') but a *negative* one ('how much harm and damage can we afford to inflict?'). Behind this formulation lies a certain view of the relationship between humanity and nature, and the philosophy (still prevalent, though no longer stridently voiced) of the 'conquest of nature'. This is the legacy of the Western legend of Prometheus who is said to have brought fire to earth in defiance of the Gods. Under the influence of that legacy, we are driven by technological hubris to undertake the 'harnessing' of nature for 'development'. That equestrian metaphor is an indicator of an essentially adversarial relationship to nature. In contrast, we have the Indian legend of Bhagiratha who brought water—the river Ganga—to earth in a prayerful spirit. In Indian tradition rivers have been regarded as deities to be worshipped, not as horses to be 'harnessed' and ridden. (That of course has not prevented us from polluting and contaminating our sacred rivers!)

If by 'sustainability' we mean the long-term maintenance of an ecological balance and thus the survival of planet earth and with it humanity; if we approach this in a positive spirit of fostering a harmonious relationship with nature rather than merely limiting the harm that we do; if we think of rivers as sacred sources; if we think of them not as separate entities but as integral parts of larger ecological systems; then our 'planning' and our ideas of 'development' will have to undergo a complete transformation.

The Human Dilemma

The fourth and final dilemma is the human rights dilemma or simply the human dilemma. The projected benefits of a project have a cost: not merely a financial cost and an environmental cost, but also a human cost; and while the benefits accrue to one set of people, the 'social costs'— a euphemism for suffering—are imposed on another and usually poorer set of people. This has been gone into in the chapter on the large-dam controversy. It must be noted that there is no necessary or inescapable dilemma here. There is no reason why the people likely to be adversely affected by a project cannot in fact be treated as 'partners in development' and made to benefit from the project; but the sad fact is that this rarely happens. A draft national rehabilitation policy has been under consideration for over 15 years, but despite repeated revisions (doubtless to make it blander and blander), it has not yet received the official imprimatur.

'Projects of the Last Resort'?

Having set forth those dilemmas, we must look for answers. One possible answer would be to say 'No more big projects'. Many do say this. That answer seems to get rid of all the dilemmas at one stroke: no need for huge investments, no induction of the corporate private sector, no environmental impacts and no displacement. (The market dilemma will not entirely disappear, but let us ignore that for the present.) However, that simple-seeming recommendation, if carefully thought through, will take us into deep waters, as we saw earlier (Chapter 11, section III). Dams, industrial projects, and so on, are manifestations of a certain conception of 'development' and a related attitude to nature. These are so well entrenched that it is very difficult for us to think on other lines. Is it possible to change them? All of us are familiar with Mahatma Gandhi's famous remark about the world having enough for everyone's need but not enough for anyone's greed. His idea of 'need' was austere; and by 'greed' he meant consumption in excess of need so defined. It seems to the author that if we are concerned about the sustainability of human life on planet earth and indeed that of the planet itself, the sanest voice that we can listen to is that of Gandhi. However, who pays heed to Gandhi now? What we call 'development' is 'greed' in his sense, and with the triumph of the 'market forces' ideology, the march of globalization and the irresistible tide of consumerism all over the world, greed seems to have come to stay. At the moment it is difficult to see how a change of direction is going to be brought about and doom averted. The advocacy of such a course is likely to be dismissed as naïve and romantic.

   If realism consists in accepting the world that we live in as a given, and if radical change seems unlikely, then perhaps we shall have to be content with the limited, flawed and as yet unimplemented concept of 'sustainable development' as generally understood, which implies the postponement rather than the elimination of doom. Even this will call for a moderation of lifestyles from what we might describe as American standards of living (using America as a symbol), not perhaps to Gandhian austerity, but to a level of modest prosperity: a difficult enough task, and one that the world as a whole cannot achieve unless America and the West in general are willing to moderate their own lifestyles.

Keeping 'sustainable development' so defined in view, we should perhaps consider what *can* be done in practical terms and in limited contexts. We may not be able to say 'No' to big-dam projects, but we can minimize their number by treating them as projects of the last resort. Such an approach would greatly reduce the severity of the dilemmas mentioned earlier, even if it does not eliminate them altogether.

That seems to bring the discussion to an end, but some difficult questions remain:

  (i) What does the expression 'projects of the last resort' mean in practical terms? Can we really cut down the number of big projects drastically? How then will our future needs be met?

  (ii) If some big projects still have to be undertaken, some of the dilemmas mentioned earlier will arise: how should we then deal with them?

Reviewing the Demand/Supply Projections

For answering the first set of questions, we need to examine: (*a*) whether the demand for water will really be of the magnitudes estimated by the NCIWRDP, and (*b*) whether, on the supply side, there are other alternatives or options.

Taking demand first, one can do no more than indicate the lines on which the examination needs to proceed. 'Demand' projections are generally based on current patterns of water-use with some adjustments for improvements in efficiency and resource-conservation, and on prevailing notions of 'development'. It is taken for granted that with a growing population, an increasing pace of urbanization and the processes of 'development', the demand for water must necessarily go up very sharply, and that the needed supplies must somehow be found. Is that self-evident? If in fact water is a scarce resource that is becoming scarcer because of increasing populations, and if a crisis is looming on the horizon, should not that consciousness of scarcity and impending crisis guide our planning? May one suggest that the approach common in the case of other consumer or industrial goods, of projecting demand and providing the supply through production, is inappropriate in the case of water? Here we need to start from the recognition of finite availability and learn to live with it. With that kind of reversal of approach, the 'demand' projections may undergo drastic changes.

Regrettably, even the NCIWRDP does not attempt such a radical re-examination. It assumes slow improvements in irrigation efficiency (from the existing level of 35–40 per cent to 50 per cent by 2025 and 60 per cent by 2050), and a very modest increase in yields from irrigated agriculture (from 3 tonnes per hectare in 2010 to 4 by 2050).[52] It also expects a saving of no more than 20 per cent over current rates of use in the industrial demand for water.[53] As for drinking water and domestic uses, the Commission would like to bring the norms for rural and urban areas closer, and projects an increase over the present norms, but does not ask whether these norms can be scaled down. That may seem a shocking suggestion, but the Commission itself recognizes that actual water supply varies widely from 50 litres per capita per day (LPCD) to 800 LPCD[54]. When we talk about the existing average supply of 182 and 103 LPCD respectively for Class I and Class II cities, we must not forget that those average supplies are not equitably distributed: there is minimal consumption by the poorer sections, fairly heavy use by the middle classes and profligate use by the affluent. Should we aim at *improving the norms for supply* from the current figures to an average of 200 LPCD for urban areas and 150 LPCD for rural areas as proposed by the NCIWRDP, or at maintaining or even reducing current norms[55] and *enforcing economies on the middle and upper classes, whether rural or urban?* Doubtless the latter would be an extremely difficult undertaking. However, should we attempt this, or should we forthwith seek supply-side solutions? Again, what will the demand projections look like if irrigation efficiency improves from the current level of 35–40 per cent

---

52 Report, pp. 57, 58.
53 Report, p. 62.
54 Report, p. 60.
55 Peter Gleick puts the basic water requirement for human needs (drinking, sanitation services, bathing, cooking and other kitchen use) at 50 litres per person per day (Gleick 1996).

to say 65 or 70 per cent instead of the 60 per cent assumed by the NCIWRDP, and earlier than the year 2050? Is that utterly impossible? Further, should we be content with a yield of 4 tonnes per hectare from irrigated agriculture? Is it possible to raise this to 6 or 7 tonnes? Should we not insist on the multiple recycling of water by industry, allowing no more than 10 per cent make-up water? Should we not strenuously promote improvements in efficiency and technological innovations in every kind of water-use to maximize what we get out of each drop of water? Apart from minimizing waste, should we not recognize that domestic and municipal waste is also a source from which water for some uses needs to be extracted? And if we do all this, will the demand picture remain the same?

As mentioned in Chapter 24, a radical re-examination of the 'future needs' is called for, and as concluded in that chapter, there *are* no easy answers to the conundrum posed.

Turning to the supply side, large-dam projects are not the only answer; alternatives and options are indeed available. It is necessary to clarify what is meant by 'alternatives'. It is not being suggested that from the perspective of national planning large-dam projects on the one hand and some other possibilities on the other can be regarded as alternative routes between which we must choose in an 'either/or' sense. Both routes may need to be used. However, in a given case there may be a choice of going in for a big dam or preferring other possibilities; and a number of choices of that kind, exercised in several cases, may mean that fewer dams are undertaken. That was what was meant by the expression 'projects of the last resort'.

We must also shake ourselves free of the usual engineering conventions of defining 'available water resources' in terms of flows in rivers, and 'usable water resources' in terms of what is stored behind a dam. What is available in nature is rainfall, not just river-flows; and while storing river waters behind a dam doubtless converts 'available' water into 'usable' water, so does *in situ* rainwater harvesting (i.e., 'catching the raindrop as it falls') and local watershed development. These are also part of the supply-side answers to the demand. (The Chopra/Goldar study does take this into account.)

Fortunately, there are many successful examples of such initiatives. Achievements such as those of Anna Hazare (Ralegan Siddhi village in Maharashtra), the late P.R. Mishra (Sukhomajri in Haryana), Rajendra Singh and the NGO Tarun Bharat Sangh (several hundred villages in

Rajasthan), and governmental efforts on a much larger scale in Madhya Pradesh under the leadership of its Chief Minister Digvijay Singh, have become well known. These are not 'small' instances but significant developments. If these examples could be replicated in the thousands across the country (wherever feasible), they could be far more significant components in national water planning than we can now imagine. A veritable transformation of the water scene may result.

There is no need to be hypnotized by visions of long-distance water transfers, whether inter-basin or intra-basin, and then plunge into despair when these turn out to be mirages. We need to look at the various drought-prone and arid areas of the country and in each case explore local possibilities of water-harvesting and conservation, keeping in mind what has been achieved in some places (e.g., Rajasthan, Madhya Pradesh, Gujarat), and thinking of recourse to external water only as a last resort. The same approach applies to urban centres. They should not be helplessly dependent on water from distant sources but must seriously explore possibilities of local augmentation.

In the case of electric power also, all options including demand management, energy-saving, increasing output from capacities already installed, minimizing energy inputs through technological improvements and innovations, extensive decentralized generation through biomass (integrating agriculture), wind/solar/tidal energy, etc., need to be explored, minimizing the need for large centralized generation. (As pointed out in the chapter on the large-dam controversy, the approaches advocated by Dr A.K.N. Reddy, Girish Sant and K.R. Datye[56] deserve more careful consideration than they have so far received.)

It must be mentioned in passing that even small and modest water-harvesting or watershed development activities are not without their effects on the environment (whether benign or adverse). These will probably be relatively minor, but they do have to be looked at. Similarly, even such local, limited initiatives could lead to conflicts, though they may not be as bitter and divisive as the conflicts generated by big projects and may be more amenable to resolution; be that as it may, mechanisms for the resolution of conflicts must be worked out and made part of the institutional arrangements relating to the initiatives.

---

56 See footnote 11.

Needed Projects: Improving Planning and Implementation

How should we proceed with the planning and implementation of those large projects that are still felt to be necessary? Not much needs to be said on the techno-economic aspects. All of us know the ills that have plagued such projects in the past: poor planning; slipshod, unprofessional processes of evaluation and investment decision; endless 'time and cost overruns'; the perpetual 'spillover' problem, with several projects continuing from Plan to Plan and never reaching closure; a thin spreading of limited resources over far too many projects; the absence of any post-project evaluation; the non-generation of revenues even to cover the running costs (operation and maintenance), much less capital-related costs or further investments, and so on. It is clear enough that we need far more professional planning; that it should be interdisciplinary, integrated and holistic, and should include a consideration of options and alternatives; that the processes of examination and investment decision should be far more stringent; that there should be a willingness to say 'No' to those projects that fail to meet the prescribed requirements; that in addition to an economic internal rate of return, a financial return criterion should be added; that there should be machinery for proper monitoring and for post-completion evaluation, and so on. All this is familiar ground. Many committees and commissions have already said these things. We can only hope that the future will be better than the past in this regard. However, something more needs to be said regarding the environmental and human concerns, which even now receive only grudging acceptance as externally imposed constraints.

Making the EIA Independent and Professional

Assuming that the prescription that project planning should be interdisciplinary, integrated and holistic is followed, what is needed thereafter is a thoroughly reliable, rigorous environmental impact assessment (EIA) as part of the investment decision processes. We saw earlier why EIAs are often unsatisfactory. The answer is to make them truly independent both of the people preparing the projects and of those approving them. EIAs should be made a professional undertaking akin to the practice of law or medicine or audit, with a statutory charter under the

Environment Protection Act, a professional council, codes of conduct, disbarment provisions and so on. Such institutional arrangements are no guarantors of quality or integrity, as we know from our experience in relation to the other professions mentioned; but they will mark a great advance over the present situation.

There should also be an independent, statutory Environmental Regulatory Authority established under the Environment Protection Act. Many of the functions now performed by the Ministry of Environment and Forests should be transferred to the Regulatory Authority.

What is particularly important is that the project proposers should not be the paymasters for EIAs; the payments should be made by the Regulatory Authority (or an office established under it) and recovered from the project sponsors, or alternatively, the latter may be required to pay on the recommendations of the former. A further possibility is to provide that the EIA agency in each case will be nominated by the Regulatory Authority from panels maintained by it. (This would be similar to the maintenance of panels of auditors by the Comptroller and Auditor General for the purpose of the audit of public enterprises.)

All this needs to be worked out in detail, but it is hoped that the approach outlined is clear enough. It is of course necessary to ensure that the regulatory system does not degenerate into one of bureaucratic control.

(A parenthesis: institutional mechanisms are only as good as the people who are appointed to them. They can be co-opted or subverted through calculated manning. One fears that this is happening already to the existing machinery on which the Supreme Court, in the Narmada judgement, placed heavy reliance. Alas, there are no remedies for this.)

The Human Aspect: True Participation/Consultation

For such projects public hearings have now become a statutory requirement under the Environment Protection Act, but this has not become fully effective yet. When it does, the affected people will undoubtedly make use of the opportunity to make their concerns known. 'People's participation' and 'consultation' of affected people are principles subscribed to by all, and are part of the prevailing policies and guidelines, but these practices do not come easily to project planners and implementers. Attention is invited to the discussion of this subject (and of the notion of 'stakeholders') in the chapter on 'Water and Rights' (Chapter 8, section II).

In regard to both the environmental and the human aspects, there are very useful recommendations in the Report of the World Commission on Dams (WCD), but the Ministry of Water Resources has compre-hensively rejected those recommendations. (See Chapter 15, section III) They feel that the requirements are too stringent and that no project will go through if they are adopted; in particular, they are very apprehensive of the principle of 'free, informed prior consent' by the affected people and the 'rights and risks' approach that the WCD has recommended.

Two comments may be made. The first is that a degree of difficulty and stringency is necessary and salutary. Such massive investments involving major interventions in nature and severe impacts on people ought *not* to have an easy passage through the appraisal and approval procedures. Few will deny that in the past there has been an absence of stringency and rigour (in evaluation) and of sensitivity and humanity (in dealing with people); this needs to change. Secondly, much of what the WCD has recommended is already part of our own policies and guidelines; and where specific procedural recommendations made by the Commission are likely to cause practical difficulties, suitable adjustments can be made. The outright and summary rejection of the WCD's recommendations was a seriously retrogressive step, because it implied the abandonment of our own principles. Efforts need to be made to rectify that grievous error. If the Ministry of Water Resources and other Ministries concerned at the Centre, the State Governments, the Indian engineering community and their professional *confrères* abroad could be persuaded to look upon the WCD's Report with less jaundiced eyes, they will find much wise counsel in it.

Conclusion

Summing up, the dilemmas that we started with can be escaped, or at least made more manageable, by the following steps:[57]

---

57 In this summing up I have not referred to the water markets dilemma, because the formulation of that dilemma earlier in this chapter is also a statement of the solution.

- reverse the usual approach of proceeding from projections of demand to supply-side answers in the form of 'water-development' projects; start with a recognition of finite supply and learn to live with it; shift the focus from 'water resources development' to 'water resources management';
- on the supply side, reverse the ranking of big 'WRD' projects as primary and local water-harvesting and watershed-development programmes as secondary and supplementary; treat the latter as primary and the former as projects of the last resort;
- where a big project is put forward as necessary, make the planning interdisciplinary from the start, with all environmental, ecological and human concerns fully internalized; and put it through a stringent process of comprehensive and rigorous evaluation to make sure: (*a*) that in itself it is a good proposition, and (*b*) that in comparative terms it represents the best choice out of the available options and alternatives;[58]
- make the EIA a truly independent and professional activity backed by a statute, a professional code, disbarment provisions, etc., under the supervision of a statutory Environmental Regulatory Authority; distance the EIA (and payments for it) from the project planners and approvers;
- transform the cost-benefit analysis into a careful, comprehensive and sensitive multi-criteria analysis;
- give primacy to the affected people, make them part of the planning and decision-making process from the start, and give them the first rights over the benefits of the project; and
- adopt the 'rights and risks' approach[59] and the principle of 'free, informed prior consent' recommended by the WCD, with such practical modifications as may be found necessary.[60]

---

58 In such a comparison the minimization of environmental and human impacts should be an important consideration; it may be useful to include 'least displacement' and 'minimum environmental impact' among the criteria for selection.

59 WCD suggests that we should ask ourselves 'Whose rights are affected?' and 'Who bears the risks?'

60 In addition to the rights and risks of the people who are affected, we need to consider also the rights of flora and fauna, of nature in general and of the river itself. The word 'rights' is of course a figure of speech here, as these mute

The approach and procedures suggested above will of course apply to future decisions, but there is an existing backlog of completed and current projects where the environmental and human impacts have not been fully recognized or properly dealt with. The clearance of that backlog of deficiencies should be the first priority, and should take precedence over new project decisions.

entities cannot speak for themselves or give their 'free, informed prior consent'. Those concerned with environmental issues will have to speak for them in the processes of environmental appraisal and clearance.

26

Linking of Rivers: Vision or Mirage?

Preliminary

The book was to have ended with the preceding chapter. The present chapter, which was not initially planned, has been necessitated by certain recent developments. The idea of the 'linking of rivers', dormant for a long time, has acquired new prominence now, particularly in the context of the acute form that the Cauvery dispute has taken, as well as the prevailing drought in several parts of the country. In response to a public interest litigation (PIL), the Supreme Court has decreed that the rivers of India shall be linked within ten years, and has directed that a task force should be set up for working out the modalities. The Prime Minister has announced the setting up of a task force and has declared that this task will be taken up on a war-footing. The leader of the Opposition has welcomed this undertaking. Three questions arise: first, whether the Supreme Court was right in issuing this direction; secondly, why the Government responded with such alacrity and enthusiasm; and thirdly, whether the idea of 'linking of rivers' (or 'inter-basin transfers' as it is sometimes referred to) is a sound one on merits.

Judicial Activism or Error?

First, is this a legitimate venture in what has come to be known as judicial activism? Please note that the question posed is not whether judicial activism is legitimate, but whether this particular direction is a legitimate exercise of judicial activism. (To obviate misunderstanding it may be stated that the author shares, with some reservations, the prevailing admiration in this country for judicial activism.)

Generally, when the judiciary stretches its scope or jurisdiction or concerns, the objective is to secure human rights or ensure justice or protect the environment; and even that last-mentioned objective can be regarded as an attempt to ensure the human right to a clean environment. No such justification is available in the present case. The Supreme Court can hold that the right to drinking water is part of the right to life, and can direct the state to ensure that that right is not denied; but precisely how that right is to be ensured is not within the domain of the judiciary. There are many different ways in which the future drinking water needs of the people can be met, and the linking of rivers is only one of the ideas mooted in this context. The Supreme Court could have directed the state to take steps to see that the need is met without specifying the particular route to be chosen for this purpose.

Moreover, it is by no means clear that there is a direct link between the right to water and the linking of rivers. The 'human right' to water is invoked in the context of water as life-support, i.e., drinking water. Drinking water is only a small part of total water needs. The really large demands for water usually arise in the context of irrigation which accounts for upwards of 80 per cent of our usable water resources. It is for meeting those huge demands that big projects—large dams, long-distance water transfers, the linking of rivers—are mooted. Thus, the link with human rights that justifies judicial activism cannot be invoked in aid of a direction for the linking of rivers. (Even if we assume that the right to food implies a right to water for irrigation—a questionable assumption—this does not necessarily translate into a right to the linking of rivers.)

It could be argued that the demand for irrigation water leads to conflicts over river waters and that the judiciary is concerned with conflicts and their prevention; but here again, the judiciary is only entitled to say: 'Find ways and means of avoiding conflicts over river waters', and not: 'Transfer waters from surplus to deficit rivers for augmenting the flows of the latter and obviating conflicts'. (A further point is that even assuming that such a transfer may help in obviating conflicts in relation to the recipient river, it may in fact generate a conflict in relation to the river from which the transfer is to be effected.)

A form that judicial activism has taken in this country is the assumption of the right to ask public authorities why they have not been discharging their responsibilities. The present case cannot be brought under that umbrella either. The Supreme Court seems to have assumed that the linking of rivers was an accepted idea that has been languishing for decades for want of attention and action. If so, a direction to accelerate

MAP 26.1: PROPOSED PENINSULAR LINKS

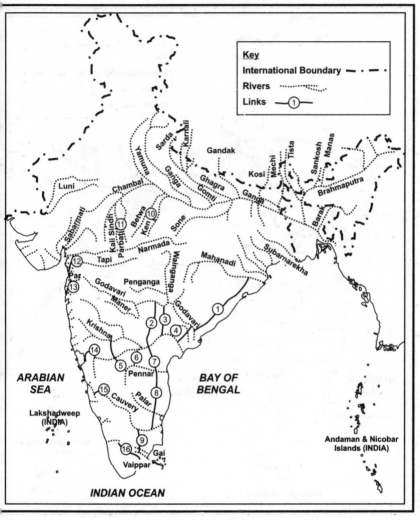

NAME OF THE LINKS

1. Mahanadi (Manibhadra)–Godavari (Dowlaiswaram)
2. Godavari (Inchampalli)–Krishna (Nagarjunasagar)
3. Godavari (Inchampalli Low Dam)–Krishna (Nagarjunasagar Tail Pond)
4. Godavari (Polavaram)–Krishna (Vijayawada)

5. Krishna (Almatti)–Pennar	6. Krishna (Srisailam)–Pennar
7. Krishna (Nagarjunasagar)–Pennar (Somasila)	8. Pennar (Somasila)–Cauvery (Grand Anicut)
9. Cauvery (Kattalai)–Vaigai–Gundar	10. Ken–Betwa
11. Parbati–Kalisindh–Chambal	12. Par–Tapi–Narmada
13. Damanganga–Pinjal	14. Bedti–Varda
15. Netravati–Hemavati	16. Pamba–Achankovil–Vaippar

Source: NCIWRDP Report

MAP 26.2: PROPOSED HIMALAYAN LINKS

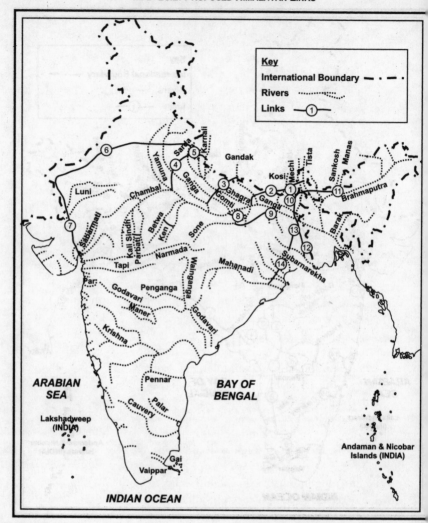

Key
International Boundary — · — · —
Rivers ·········
Links —①—

ARABIAN SEA

Lakshadweep (INDIA)

BAY OF BENGAL

Andaman & Nicobar Islands (INDIA)

INDIAN OCEAN

NAME OF THE LINKS

1. Kosi–Mechi
2. Kosi–Ghagra
3. Gandak–Ganga
4. Ghagra–Yamuna
5. Sarda–Yamuna
6. Yamuna–Rajasthan
7. Rajasthan–Sabarmati
8. Chunar–Sone Barrage
9. Sone Dam–Southern Tributaries of Ganga
10. Brahmaputra–Ganga (MSTG)
11. Brahmaputra–Ganga (JTF) (ALT)
12. Farakka–Sunderbans
13. Ganga–Damodar–Subarnarekha
14. Subarnarekha–Mahanadi

Source: NCIWRDP Report

action may seem a legitimate exercise of judicial activism. That is not the case. The linking of rivers did not figure in the Ninth Plan, and so far as one knows, the Tenth Plan document now awaiting release does not mention it; a reference may of course be added now. The Report of the National Commission for Integrated Water Resources Development Plan (NCIWRDP) 1999 did not make any major recommendations in this regard; we shall return to this. The Prime Minister's address to the National Water Resources Council (NWRC) commending the National Water Policy (NWP) 2002 for adoption did not refer to it. Thus it is clear that the Government did not have any river-linking project on the anvil. The idea is doubtless an old one, but there have always been doubts about its soundness and practicability.

All these points lead to the conclusion that the Supreme Court's direction in this case was not at all a defensible instance of judicial activism. One wonders whether it was a *judicial* act at all. Be that as it may, the direction has clearly foisted a new decision on the Government, given it a priority and virtually mandated an allocation of funds.

Further, the judicial direction has been given in advance of the usual examination and clearances (in fact, in advance even of the formulation of a project or projects!). One wonders whether this does not constitute a violation of the prescribed procedures, not to mention the statutes with which the Ministry of Environment and Forests and other Ministries are concerned. Given the judicial direction and the Prime Minister's statement, can those Ministries and the Technical Advisory Committee be expected to examine the project(s) in question objectively and rigorously? Lastly, and this is a very important point, the citizen usually has the right to move the courts with writ petitions against executive action on certain grounds, but that possibility seems to be ruled out in a case in which executive action is taken in pursuance of a judicial direction. It is ironic that the Supreme Court, which is usually anxious to assert its power of judicial review, has lost it in this case by becoming a party to executive action.

The Government's Response

The Government of India is bound to act on a direction of the Supreme Court, but it is interesting to observe that it has done so with uncharacteristic promptitude and enthusiasm. The reasons for this are not far to seek. The Supreme Court has presented the ruling party with a politically attractive proposition, and that party has been quick to adopt it and make fervent declarations: it evidently hopes to extract considerable political advantage out of this dramatic project (or clutch of projects). The Opposition, for its part, cannot afford to be seen as opposing an idea that seems to be in the national interest, and has been obliged to welcome it.

That is the political dimension; there is also a bureaucratic angle. The Ministry of Water Resources (MoWR) at the Centre has for long been trying to enlarge its role, but has been finding this difficult because of resistance by the State bureaucracies. Reference to this tug-of-war was made in the discussion on the NWP 2002 in an earlier chapter. The MoWR has been arguing the case for the transfer of 'water' to the Concurrent List, but with little prospect of success. Against that background, the Supreme Court's direction on the linking of rivers must have been very welcome to it, because any such large national undertaking on inter-State rivers is bound to enlarge its role substantially. Not only is its 'clout' vis-à-vis the State bureaucracies likely to increase, but its relative importance among the Ministries at the Centre may also go up. Its position will be further strengthened if there is new legislation to underpin the river-linking idea.

Further, it is interesting to note that the Prime Minister, who gave a resounding call at the last meeting of the NWRC for a national campaign on rainwater harvesting and for the recognition of the community as the custodian of water resources, has not set up any task force to promote those ideas, but has done so promptly on the linking of rivers, and that there is considerable excitement in governmental circles over this idea. Gigantism always casts an irresistible spell on our bureaucracy and technocracy as well as on our politicians.

(One is dismayed at the thought of the enormous 'opportunities' that public expenditures of the magnitude involved will present to certain elements in the bureaucracy/technocracy and the political class; but that is another story.)

Examination of the Idea

Let us turn now to a consideration of the merits of the proposition. The notion of the linking of the rivers in the subcontinent is an old one. In the 19th century, Sir Arthur Cotton had thought of a plan to link rivers in southern India for inland navigation. The idea was partially implemented but was later abandoned because inland navigation lost ground to the railways. Even the canal that was constructed went into decline.

A phrase that caught the imagination of the people and passed into popular parlance was 'Garland Canal'. This idea (which was not quite the same as the linking of rivers) was mooted by Captain Dinshaw J. Dastur, an air pilot. It was merely a fanciful notion that never commanded respect among knowledgeable people. The catchy phrase refuses to die and keeps surfacing from time to time, but does not merit serious discussion here.

An idea that has exercised the minds of the Indian water resource planners for a long time is that of tapping the surplus resources of the mighty Brahmaputra. A significant part of the water resources of India, estimated in terms of the flows near the terminal points of the river systems, lies in the Brahmaputra, which, unfortunately, is in a remote corner of the country, far from the areas where the demand for water is high. There has therefore been a preoccupation with the idea of a transfer of water from that river to places where it is needed. In the talks with Bangladesh over river waters in the 1970s, India proposed a gigantic (100,000 cusec) Brahmaputra–Ganga gravity link canal taking off from Jogighopa in India, passing through Bangladesh, and joining the Ganga just above Farakka. The proposal was rejected by Bangladesh for many reasons, at least some of which were and continue to be valid; that scheme is virtually dead. An alternative link canal passing entirely through Indian territory (the Siliguri chicken-neck!) will involve large lifts and seems likely to be both non-viable and questionable from other points of view, even if it is physically feasible and the money can be found. The idea has not been seriously pursued, and for good reason. We must disabuse ourselves of the notion that the vast waters of the Brahmaputra can be diverted westwards or southwards. At best we can think in terms of some minor transfers within the eastern region.

Dr K.L. Rao's proposal of a Ganga–Cauvery Link was another idea that (like Captain Dastur's 'Garland Canal') appealed to the general public and acquired an enduring life. As envisaged by Dr Rao, the link was to take off near Patna, pass through the basins of the Sone, Narmada, Tapi, Godavari, Krishna and Pennar rivers, and join the Cauvery upstream of the Grand Anicut. It was to have been 2,640 kilometres long, withdrawn 60,000 cusec from the flood flows of the Ganga for about 150 days in the year, and involved a lift of a substantial part of that water over 450 metres. The scheme was examined and found impractical because of the huge financial costs and the very large energy requirements. However, the idea survives in the popular mind and comes up whenever water scarcity is felt and conflicts (such as the Cauvery dispute) become acute in the southern parts.

Apart from considerations of techno-economic viability, on which the proposition was abandoned, the diversion of waters from the Ganga will have international implications. Bangladesh is likely to view this with apprehensions and raise objections. This needs to be gone into briefly. Under the India–Bangladesh Treaty of December 1996 on the sharing of Ganga waters, India has undertaken to protect the flows arriving at Farakka, which is the sharing point. Bangladesh may contend (rightly or wrongly) that a diversion of waters from the Ganga to the southern rivers will not be consistent with that undertaking. Besides, it is a proposition accepted by both India and Bangladesh that the Ganga is water-short in the lean season and needs to be 'augmented', though the two sides have different notions on the means of augmentation: that is a debatable proposition, but if that is in fact true, where is the scope for diversion from the Ganga? India may argue that only the flood flows of the Ganga will be stored and diverted, and that the lean season flows (which are what Bangladesh is concerned with under the Treaty) will not be affected; but Bangladesh would say that if the flood flows can be stored, the stored waters should be used for the augmentation of the lean season flows of the Ganga itself for being shared at Farakka, and not diverted to other basins. (India for its part had earlier taken the position that Farakka was too far away to receive any significant augmentation from storages in the Ganga system in the Himalayan region, which is where storages are considered feasible; it may be difficult to reconcile that argument with proposals for large diversions to the southern rivers.) Within India, Bihar has already a strong sense of grievance that its interests in respect of the waters of the Ganga system have not been given due consideration; and West Bengal has only reluctantly agreed to the large allocations to Bangladesh under the Ganga Treaty and has been pressing the needs of Calcutta Port. Neither State will look kindly upon any diversion of Ganga waters southwards.

Having ruled out the idea of a Ganga–Cauvery link as unworkable, the Ministry of Water Resources (or whatever it was called then) brought out a booklet on the National Perspective for Water Development in August 1980. In pursuance of the perspectives set forth in that booklet, the National Water Development Agency (NWDA) was established in 1982 for working out basinwise surpluses and deficits and studying the possibilities of storages, links and transfers. During the past two decades the NWDA has done a great deal of work and produced an impressive amount of documentation. It undertook the studies in two main components, namely the Himalayan rivers component and the Peninsular rivers component.

The Himalayan component envisages a number of links that need not be detailed here. The general idea is to transfer water from the Brahmaputra and Ganga systems westwards to southern Uttar Pradesh, Haryana, Punjab and Rajasthan, and perhaps eventually southwards to the peninsular component. As data relating to the Himalayan rivers are classified as confidential, even the NCIWRDP found it difficult to study these proposals. The Commission merely observed that the costs involved and the environmental problems would be enormous; that the further expansion of irrigation in the desert areas of Rajasthan would need examination from all angles; that the NWDA's Himalayan component would require more detailed study and that the actual implementation was unlikely to be undertaken in the immediate coming decades (Report, September 1999). To those cautionary remarks of the NCIWRDP it may be added that in so far as transfers from Manas, Sankosh, Karnali, etc., are involved, Bhutan and Nepal would need to be consulted, and as already mentioned, Bangladesh would have something to say in so far as the waters of the Brahmaputra and the Ganga are concerned.

The Peninsular rivers component involves a number of links, of which the most important would be those connecting Mahanadi, Godavari, Krishna, Pennar and Cauvery. The details are given in Table 26.1.

Other links not included in the above Table would include Ken–Betwa, Parbati–Kalisindh–Chambal, Par–Tapi–Narmada, Damanganga–Pinjal, etc. Another idea is the partial diversion of certain rivers flowing into the Arabian Sea eastwards to link with rivers flowing into the Bay of Bengal (Bedti–Varda, Netravati–Hemavati, Pamba–Achankovil–Vaippar).

All these proposals were studied by the NCIWRDP. After a careful examination of the water balances of the various basins, the Commission observed: 'Thus there seems to be no imperative necessity for massive water transfers. The assessed needs of the basins could be met

Table 26.1

S. No.	Name of Link	From River	To River	Annual Volume of Transfer (mm³)
1.	Manibhadra to Dowleswaram	Mahanadi	Godavari	11,176 (6,500)*
2.	Inchampalli to Nagarjunasagar	Godavari	Krishna	16,426 (14,200)
3.	Inchampalli to Pulichintala	Godavari	Krishna	4,371
4.	Polavaram to Vijayawada	Godavari	Krishna	4,903 (3,305)
5.	Almatti to Pennar	Krishna	Pennar	1,980
6.	Srisailam to Pennar	Krishna	Pennar	2,310 (2,095)
7.	Nagarjunasagar to Somasila	Krishna	Pennar	12,146 (8,648)
8.	Somasila to Grand Anicut	Pennar	Cauvery	8,565 (3,855)
9.	Kattalai Regulator to Vaigai to Gundar	Cauvery	Vaigai	2,252

Source: NCIWRDP's Report, September 1999.
*In column 5, the upper figure indicates the gross diversion while the lower figure in brackets gives the quantity reaching the recipient river. The difference is explained by utilization and losses en route.

from full development and efficient utilization of intra-basin resources except in the case of Cauvery and Vaigai basins. Therefore, it is felt that limited water transfer from Godavari at Inchampalli and Polavaram towards the south would take care of the deficit in Cauvery and Vaigai basins ... Though surplus is available in Mahanadi also, the transfer from that river would require much longer link and is in any case not required for the immediate future ...' (The Commission then takes note of some uncertainties that may affect the above judgement, and says that further studies as to the future possibilities of inter-basin transfers need to be continued.)

Three points need to be noted here. First, the assessment that surpluses are available in Mahanadi and Godavari is not shared by the

Orissa and Andhra Pradesh Governments. Secondly, there is considerable opposition to the idea of the eastward diversion of west-flowing rivers. Thirdly, there is irony in the proposition that the answer to the difficulty of persuading Karnataka to release Cauvery waters for Tamil Nadu (a co-riparian State) lies in the even more difficult course of persuading Orissa to spare Mahanadi waters for non-riparian States! We are ready to project a shortage in a basin and draw the conclusion that water must be brought from another basin. In reality, the answer to the sharing problem in the Cauvery lies in both Tamil Nadu and Karnataka learning to reduce their excessive demands on the waters of the river through a combination of measures: the 'shortage' will then disappear.

Let us turn from those specifics to theoretical considerations. We cannot simultaneously urge: (i) that planning must be on the basis of a basin as a natural hydrological unit, and (ii) that we must cut across the basins and link them. Quite apart from the technical challenges involved, this implies the redrawing of the geography of the country. One's misgivings about that kind of technological hubris or Prometheanism ('the conquest of nature' philosophy) may be dismissed by some as romantic, but the practical difficulties involved cannot be so dismissed. Barring a few cases where short gravity links may be feasible, inter-basin transfers generally involve the carrying of water across the natural barrier between basins (which is what makes them basins) by lifting, or by tunneling through, or by a long circuitous routing around the mountains if such a possibility exists in a given case. Rivers or streams may also have to be crossed in some cases. Exceptionally heavy capital investments and continuing energy costs (in operation) are almost always likely to be involved. In addition, big dams, reservoirs and conveyance systems will need to be built, involving substantial environmental impacts and displacement/rehabilitation problems. All this will need to be looked at very closely in every case. Thorough Environmental Impact Assessments, Cost-Benefit (multi-criteria) Analyses, qualitative assessments of non-quantifiable considerations, and based on these, rigorous investment appraisals, will need to be undertaken. Not too many projects are likely to survive such a scrutiny.

Even more serious is the funding problem. Plan outlays are barely adequate even for the completion of projects already undertaken. One estimate—that of the NCIWRDP—of amounts needed for completing spillover projects is Rs 70,000 crore in the Tenth Plan and Rs 1,10,000 crore in the Eleventh Plan. That leaves no scope for new major projects, and necessitates a severe selectivity even in regard to the continuance of what are called 'on-going projects'. Against that background, there seems to be little likelihood of finding the massive resources needed for a major river-linking undertaking. The rough figure mentioned in the Supreme Court in this context was Rs 5,60,000 crore! The popular expression 'one does not know whether to laugh or to cry' comes to mind. One is reminded of the Tamil saying 'Asking for directions to a place to which one is not going'. We may be wasting a good deal of time in pursuing such grandiose and unpromising propositions, and distracting ourselves from finding time and money for more modest, worthwhile and urgent activities, such as extensive water-harvesting all over the country (wherever feasible) and the massive task of rehabilitation of tanks in the South and other similar traditional systems dying wisdom elsewhere. Even more important is effective demand management through improved efficiency and economy in water-use, whether in agriculture or in industry or in domestic and municipal uses, so as to minimize the need for supply-side solutions.

We must hope that the Task Force set up as directed by the Supreme Court will consider not merely the 'modalities' of the 'linking of rivers' but also the soundness and wisdom of the idea. Any headlong rush in the pursuit of this chimera will be disastrous.

Conclusion

Having examined the grand vision of the linking of rivers and found it more of a mirage, what then must we do? The answer is clear. We are driven back to the plan of action outlined at the end of the preceding chapter. The conclusion of that chapter is also the conclusion of this chapter and indeed of the book.

Appendix I

List of Articles and Papers by the Author which have been used to Varying Degrees in this Book

(Note: Papers not significantly used or drawn upon have not been listed. Book reviews have also not been mentioned.)

Papers/Chapters Contributed to Books Edited by Others:

'Dispute and Resolution: The Ganges Treaty', in *Indian Foreign Policy: Agenda for the 21st Century*, Vol. 1 & 2, Foreign Service Institute, in association with Konark Publishers Pvt. Ltd., New Delhi, 1997).

'Water Resources Planning: Changing Perspectives' in *Environmental Management: Indian Perspectives*, eds. S.N. Chary and Vinod Vyasulu, Macmillan India, 2000.

An essay on the large dam controversy in the section on 'Indian Dams' in *History in Dispute, Vol. 7, Water and the Environment Since 1945: Global Perspectives*, eds. Char Miller, Mark Cioc, Kate Showers, Detroit: St. James Press/Gale Publications, 2001.

'Three River Water Treaties' in *Conflict and Peacemaking*, ed. P. Sahadevan, Lancer's Books, New Delhi, 2001.

Papers/Monographs brought out by Centre for Policy Research:

'The Cauvery Dispute', May 1996.
'Water Resource Planning: Changing Perspectives', February 1999.
'Conflict Resolution: Three River Treaties', June 1999.
'Water: Charting a Course for the Future', September 2000.

Articles Published in the *Economic and Political Weekly*, Mumbai:

'Large Dams: The Right Perspective', 30 September 1989.

'Federalism and Water Resources', 26 March 1994

'Scarce Natural Resources and the Language of Security', 16 May 1998.

'Water Resource Planning: Changing Perspectives', 12 December 1998.

'Conflict Resolution: Three River Treaties', 12 June 1999.

'The Fallacy of Augmentation', 14 August 1999.

'A Judgement of Grave Import', 4 November 2000.

'Water: Charting a Course for the Future', 31 March and 14 April 2001.

'World Commission on Dams and India: Analysis of a Relationship', 23 June 2001.

'The New National Water Policy', 4–10 May 2002.

'Was the Indus Waters Treaty in Trouble?', 22 June 2002.

'Inter-State Water Disputes Act 1956: Difficulties and Solutions', 13–19 July 2002.

'Linking of Rivers: Judicial Activism or Error?', 16 November 2002.

Other Articles/Papers

'Water Policy I—Towards Optimal Use', *The Hindustan Times*, New Delhi, 16 November 1993.

'Water Policy II—Not Much Headway', *The Hindustan Times*, New Delhi, 17 November 1993.

'Indian Federalism and Water Resources', *International Journal of Water Resources Development*, Oxford, UK, Vol. 10(2), 1994.

'The Cauvery Dispute', *Liberal Times* (Friedrich Naumann Stiftung), New Delhi, Vol. 4 No. 1, 1996.

'Towards the Resolution of the Ganga Waters Dispute', *Politics India*, New Delhi, Vol. 1, No. 4, October 1996.

'India Must Bridge the Divide on Ganga Waters', *The Times of India*, New Delhi, 10 December 1996.

'Ganga Waters Treaty: Give it a Fair Chance', *The Times of India*, 15 July 1997.

'The Ganges Treaty in Operation', *The Daily Star*, Dhaka, Bangladesh, 24 July 1997.

'The Indo–Bangladesh Ganga Waters Dispute', *South Asian Survey*, Vol. 4, No. 1, January–June 1997, New Delhi, Sage Publications India Pvt. Ltd., New Delhi.

'Inter-State River Water Disputes: Some Suggestions', *Mainstream*, 5 June 1999.

'Water–Large and Small', *The Hindu Survey of the Environment 2000*.

'Water and Security in South Asia', paper prepared for a Project on Environment/Security Linkages, Regional Centre for Strategic Studies, Colombo, and presented at a Conference at Islamabad in May 2000. (An edited version of the paper is also being published in the book *Environmental Dimensions of Human Security: Perspectives from South Asia*, University Press of America, 2002.)

'Large Dam Projects: The Framework of Laws, Policies, Institutions and Procedures', Paper prepared for the India country study for the World Commission on Dams, June 2000.

'Large Dams, Trans-Boundary Waters, Conflicts', Paper prepared for the World Commission on Dams, June 2000.

'Sustainable River Basin Management: The Case of India', Paper prepared for a Conference organized by Lanka International Forum for Sustainable Development, Sri Lanka Water Partnership, Munasinghe Institute for Development and the International Water Management Institute at Colombo on 17–19 July 2000.

'Water: Transforming Laws and Institutions', *The Asian Journal: Journal of Transport and Infrastructure*, Vol. 7, No. 3, September 2000.

'Water: A Note Regarding Conflicts', Paper prepared for Institute for Resource Management and Economic Development's (IRMED) International Conference on Socio-Economic, Institutional and Environmental Aspects of Sustainable Development of Water Resources, New Delhi, 27–30 November 2000.

'Water And Rights: Some Partial Perspectives', Paper prepared for a Conference on Water Resources, Human Rights and Governance organized by Nepal Water Conservation Federation, Kathmandu, and Institute for Social and Environmental Transition, Boulder, Colorado, USA, at Kathmandu on 26 February–2 March 2001.

'Water-Related Conflicts: Factors, Aspects, Issues', Paper presented at a Seminar on Conflict Prevention and Management in South Asia organized by the European Centre for Conflict Prevention, Utrecht, The Netherlands, Institute of Peace and Conflict Studies, New Delhi, and International Centre for Peace Initiatives, New Delhi, at New Delhi, 15–17 May 2001. (Being included in Mekenkamp, M.; P. van Tongeren P; H. van de Veen, (eds.), *Searching for Peace in Central and South Asia: An Overview of Conflict Prevention and Peacebuilding Activities*, under publication by Lynne Rienner Publishers, Boulder, London.).

'Delay and drift on the Mahakali', *Himal, South Asian Magazine*, Kathmandu, June 2001.

'Water: Commodity, Commons, Basic Right, Divinity', paper prepared for an International Conference on 'Globalization, Environment and People's Survival' (September–October 2001), and brought out under the 'Lok Swaraj Series' by Navdanya and the Research Foundation on Science, Technology and Ecology, New Delhi.

'Groundwater Legislation', Paper prepared for Workshop on Water Harvesting with Special Reference to Artificial Recharge, National Geo-Physical Research Institute, Hyderabad, 9–10 October 2001, on the occasion of the Annual General Meeting of the Geological Society of India.

'The Dilemmas of Water Resources Development', Paper prepared for the Second Biennial Conference of the Indian Society for Ecological Economics, Bhopal, 19–21 December 2001.

'Review of Studies of India's Future Water Requirements' and 'Some Clarificatory Remarks on 'Population and the Environment', prepared in December 2001–January 2002 for the Wellcome Trust India/Population Project (London School of Economics), Wellcome grant no. 053660.

'Cauvery: Disturbing Developments', *The Hindu*, 22 September 2002.

'The Cauvery Tangle: What is the Way Out?', *Frontline*, Chennai, 27 September 2002.

'Disputes Over Sharing of Inter-State River Waters–Towards Cooperation and Conflict Resolution', *The Sunday Tribune*, 6 October 2002.

'Rivers of Discord', article on the Linking of rivers, *The Times of India*, 9 November 2002.

'A Cauvery Debate', *Frontline*, 6 December 2002.

'Linking of Rivers: Vision or Mirage?', *Frontline*, 20 December 2002.

Appendix II

The Cauvery Dispute: A Debate at Bangalore

An informal discussion on the Cauvery dispute took place recently at Bangalore at my initiative. The group was a small non-official (largely professional) one. At the outset I referred to the recent developments, and stressed the need for a 'civil society' initiative to find a solution to the problem, as also to promote goodwill between the people of the two States. The discussions that followed covered many points. The present note aims at bringing those points to the notice of a wider audience. In the following paragraphs I set forth the points raised and comment on them. Needless to say, the comments reflect my views.

(1) *The Supreme Court has no jurisdiction*
(This point was made with reference to the recent and continuing proceedings in the Supreme Court.)

It is true that once a dispute on an inter-State river is referred to a Tribunal under the Inter-State Water Disputes Act 1956 (ISWD Act), the jurisdiction of the courts is barred. However, what has gone before the Supreme Court (SC) in this case is *not* the water dispute, which is still before the Tribunal, but the question of *implementation of the Tribunal's Interim Order*, and the related issue of compliance with the decisions of the Cauvery River Authority and with the directions of the Supreme Court itself. These are entirely within the SC's jurisdiction. It is evident that the SC had no doubts in this regard.

(2) *Adjudication is not the best means of settling such disputes. Mutual agreement through negotiations is the best course. Conciliation and arbitration should also have been tried*
Certainly, a negotiated settlement is the best course, but when that fails (as it did in the present case—20 years of talks produced no result), adjudication becomes necessary as a last-resort. This is what Article 262

and the ISWD Act provide for. These are very useful and necessary last-resort provisions, and important components of our federalism. (Some perceived deficiencies in the ISWD Act have also recently been sought to be set right through amendments.) It is not right to be negative and dismissive towards these provisions. There is no reason why adjudication should necessarily be divisive; it can be approached in a constructive, cooperative spirit.

The question of arbitration under the River Boards Act 1956 (RBA) will be discussed later, but in general, it may be noted that once a dispute is referred to arbitration, thereafter there is not much difference between that course and adjudication; each side argues its case strongly, but finally the arbitrators' or adjudicator's award has to be accepted.

Given goodwill and reasonableness on both sides, any route—nego-tiations, conciliation, arbitration, adjudication—can be made to work; in the absence of goodwill and reasonableness, nothing will work. In-cidentally, the recourse to adjudication does not rule out continuing negotiations or non-official efforts at conciliation. If through these means an agreement can be reached, that can be reported to the Tribunal and converted into an Award. That was what happened in the Godavari case: the Godavari Tribunal's Award is nothing but an agreement arrived at by the concerned States.

(3) *The entire action under the ISWD Act in this case is illegal. There is another Act, namely the RBA. That is a far more comprehensive Act with greater potential for constructive management. It provides for arbitration in the event of disputes. That was the route that should have been followed. The ISWD Act itself rules out adjudication under it if arbitration under the RBA is possible. A writ petition challenging the recourse to the ISWD Act in this case has been filed and is pending.*

We must await the outcome of the writ petition which is said to be pending, but evidently it has not prevented adjudication from proceed-ing and the Karnataka Government from participating in the proceed-ings. After 12 years of participation, it seems doubtful whether they will or *can* now argue that the entire proceedings were illegal.

Apart from that, the argument outlined above is based on a misun-derstanding of the relationship between the two Acts. The RBA is primarily concerned with the planning and management of inter-State rivers and river valleys and not with conflict-resolution, though it does provide for the arbitration of disputes (arising in the context of the functioning of the Board). That Act was passed under Entry 56 in the Union List which gives the Central Government a role in relation to inter-State rivers to the extent that Parliament legislates for it. The ISWD Act was enacted under an entirely different constitutional provision, namely Article 262 which deals specifically with inter-State river water-disputes and enables Parliament to pass legislation for their adjudication. The ISWD Act certainly rules out recourse to it if the route of arbitration under the RBA is available, but a prerequisite for that is the existence of a River Board. There is no Cauvery River Board.

(Incidentally, under the RBA the arbitrators are appointed by the Board, and not nominated by the parties to the dispute; and the arbitrators will be judges assisted by technical assessors. In other words, arbitration under the RBA will be exactly similar to adjudication under the ISWD Act!)

No Board has been set up under the RBA because of resistance by the States. It is a well-known fact, frequently commented upon in related literature, that the RBA has been inoperative and virtually a dead letter. When the National Water Policy 1987 was being drafted in 1985–86, the question of river basin organizations came up, but the idea was stoutly resisted by several Chief Ministers, including the then Chief Minister of Karnataka, Ramakrishna Hegde. More recently (in 1997), when the Central Government circulated a draft notification for the establishment of a Cauvery river authority in the form of a standing professional and empowered body, it was rejected by Karnataka, and eventually a purely political committee called the Cauvery River Authority (CRA) with the Prime Minister as Chairman and the Chief Ministers as Members was set up with the limited function of monitoring the implementation of the Interim Order of the Tribunal and resolving disputes in that context.

Against that background, it is clear that the alternative of arbitration *via* the RBA was not available in this case because there was no Cauvery River Board. The route of adjudication under Article 262 and the ISWD Act was very much available and was taken after prolonged talks proved fruitless. (In fact, even after Tamil Nadu asked for a Tribunal in 1986 the Centre continued to explore the possibility of a negotiated settlement for four more years; it was only under the specific direction of the SC that the Tribunal was set up in June 1990. It can hardly be argued that the SC was ordering an illegal course of action!) It seems unrealistic to suggest at this stage, when the final Order of the Tribunal seems likely to be received within a matter of months, that the work of 12 years should be abandoned and an alternative route embarked upon. That would entail first the establishment of a Cauvery River Board (assuming that Karnataka would agree to this, which is very doubtful), and then the reference of the dispute to it for arbitration. Such a suggestion cannot be seriously entertained.

It was argued that this should be the course of action in future cases. By all means let us try to bring the moribund RBA back to life and set up a number of River Boards, if that is politically feasible. However, that course of action is not available at this stage as a means of resolving the Cauvery dispute.

(4) The 1991 Interim Order of the Tribunal is patently unimplementable. How can an unvarying quantum of 205 TMC be released year after year without regard to changing circumstances?

The figure of 205 TMC was arrived at by the Tribunal by taking the average of 10 years, after eliminating the best and worst years. It must have seemed a fairly safe figure, considering that the historical use of Cauvery waters by Tamil Nadu, as determined by the Central Fact-Finding Committee of 1972–73, was much higher. Tamil Nadu has therefore to manage with less water than they were using in the past. That is as it should be. Besides, there has been no difficulty in recent years when the rains were good. The quantum of 205 TMC presents no problems except when the rains fail. The weakness of the Interim Order was that it did not provide a formula for distress-sharing in lean years. When Karnataka went back to the Tribunal in 1992 saying that the IO was unimplementable, the Tribunal reaffirmed its order, but said that in a difficult year a *pro rata* adjustment could be made. That was merely an observation. No formula was laid down. That was the source of all the subsequent trouble. The Tribunal must since then have taken note of the difficulties actually experienced, and the Tamil Nadu and Karnataka Governments must also have argued these points in their presentations to the Tribunal. It seems certain that the final order will contain: (i) a proper water-sharing pattern taking all the points urged by the parties to the dispute, and (ii) suitable provisions to cover the contingency of low-flow years.

(5) In a year of distress (like the current year), when there is not enough water even for its own farmers' needs how can Karnataka spare water for Tamil

Nadu? This will be totally unacceptable to Karnataka farmers. These ground realities must be recognized. As the water levels in the reservoirs fall, tempers will rise further. The relationship between Kannadigas and Tamils will come under further strain. The Tribunal, the CRA, and the SC do not recognize these realities

To what extent should the Karnataka Government be guided by what is acceptable to the farmers in Mysore/Mandya? The latter have (either on their own or under the advice of some leaders) taken an extreme position. They assert their exclusive rights over Cauvery waters, and have said that 'not a drop' should be released to Tamil Nadu. They hold their position with passion and mount violent agitations. They are asking the State Government to ignore the decisions of the Tribunal, the CRA, and the Supreme Court, and have virtually declared that their own decisions will override the decisions of all those authorities. (Even after the SC's severe strictures and the Karnataka Government's apology, the farmers are continuing their strident position and their agitations.) Are these the 'ground realities' that we must accept? The Karnataka farmers in the Cauvery basin have been profoundly misguided. It is the responsibility of the Karnataka Government and politicians (of all parties) to educate them on the right approach. Instead, they are being guided by the sentiments of the farmers. This is very unfortunate and fraught with serious consequences. (It is also unfortunate that the intelligentsia have been silent.) In Tamil Nadu, film stars have jumped into the fray and are holding meetings and going on fasts; that is equally deplorable.

Even in years of poor rainfall and low flows the sharing principle remains valid. There has to be a sharing of the distress. There is no question of anybody 'giving' or 'sparing' water for anyone else. No one—not Kerala or Karnataka or Tamil Nadu or Pondicherry—'owns' Cauvery waters: all have rights of use. The shares will vary according to various criteria, but there is no hierarchy of rights. In the Ganga Treaty of 1996, India has recognized that Bangladesh as the lower riparian has rights over the Ganga waters; it has gone to the extent of undertaking to protect the flows arriving at Farakka, which is the sharing point. In situations of exceptionally low flows, the two Governments have to enter into consultations on an emergent basis, but the Treaty rights of Bangladesh are not extinguished. (That Treaty was signed on behalf of India by the then Prime Minister Deve Gowda.) Unfortunately, Karnataka has not been willing to recognize that Tamil Nadu has *rights* as a lower riparian. It has been talking about 'giving' or 'sparing' waters, implicitly assuming the primacy of its own rights as the upper riparian. It reserves the right to determine how much it needs and how much it is willing to release. This is nothing but the Harmon Doctrine which commands no acceptance; neither does the doctrine of prior appropriation or prescriptive rights adopted by Tamil Nadu. What finds general acceptance, and has been adopted by successive tribunals, is the principle of equitable sharing for beneficial uses. Tamil Nadu and Karnataka have to recognize each other's rights, needs and problems: that is the only way in which this dispute can be resolved.

Purely as a water-sharing dispute this is not a particularly difficult one. (The difficulty in this case lies not in water but in politics.) Some kind of a sharing pattern was attempted by the Central Government in the 1970s and we can proceed further from there. Out of the available flows of 671 TMC (as determined by the Fact-Finding Committee in 1972–73), some 40 TMC or so have to be earmarked for Kerala and around 10 TMC for Pondicherry (these are purely hypothetical figures), leaving around 621 TMC for being shared between Karnataka and Tamil Nadu. This can be split as 416/205 or 400/221 or 380/241 or 350/271 (or some such division) depending on the factors and criteria adopted. The range within which a decision has to be taken is fairly narrow. The Tribunal will doubtless take all the relevant factors and all the arguments put forward by the four States into account and arrive at its final Order, and will surely set forth its reasoning at great length; and it will also surely lay down the manner of sharing in the event of low flows.

(6) *We do not expect a good report from the Tribunal because Karnataka has not argued its case properly before the Tribunal; many important aspects and issues have not figured in the proceedings*
It is very difficult to believe that in 12 years of proceedings the officers, engineers and lawyers of Karnataka have failed to present the State's case fully, competently and comprehensively. If indeed this is the case, the State must accept the consequences; but *prima facie* this seems very unlikely. Assuming that there are some points and arguments that have been overlooked, there are two remedies. First, even at this late stage, the possibility of making further submissions to the Tribunal can be explored. Secondly, when the Tribunal gives its Award, that is not the end of the story. Within three months, one or more of the States concerned or the Centre can (in terms of a provision in the ISWD Act) make a further reference to the Tribunal seeking clarifications or a

supplementary report; even new points can be raised at that stage. Thus, if any of the four States involved in this dispute is dissatisfied with the final order of the Tribunal, it has an opportunity of saying so to the Tribunal. The Tribunal will then consider the points raised and give clarifications or a supplementary report. It can substantially modify or revise its final Order at that stage.

(7) *Karnataka has a sense of grievance about the past. In the 1892 and 1924 Agreements the princely State of Mysore was at a disadvantage in relation to Madras Presidency which was part of British India*
This is disputed by Tamil Nadu which argues that the British Government was fair and objective; that the negotiations leading to the 1924 Agreement were long and hard, and that the Agreement was welcomed by the then Diwan of Mysore. Be that as it may, it must be recognized that Karnataka does have a sense of grievance. However, that bit of history is no longer relevant. Assuming that Karnataka did have a grievance about the 1924 Agreement, and granting that in comparison with Tamil Nadu it was a late starter on irrigated agriculture, that initial disadvantage has since been remedied to a large extent. As the upper riparian it has physical control over the waters, and has proceeded to build several dams and reservoirs on the different rivers in the Cauvery system, reducing flows into Tamil Nadu. There is no ground now for Karnataka to nurse a sense of grievance; today, Tamil Nadu as the lower riparian suffering reduced flows is the complainant.

(8) *It is hardly fair that Tamil Nadu should insist on taking three crops when Karnataka is not able to take even one. And Tamil Nadu wants to grow paddy, paddy, paddy*
Tamil Nadu has its answers to these points, but we need not go into them here. Both States are growing water-intensive crops (paddy in Tamil Nadu, sugarcane in Mandya). Both have to learn to use water better, and if that calls for changes in cropping patterns, they should go in for such changes. If water availability is restricted, as it is bound to be under any allocation of a limited quantum, farmers in both States will adjust themselves to that situation over a period of time. The Cauvery Delta area in Tamil Nadu is already getting much less water than it used to in the past, and the process of adjustment has begun. Over a period of time, farmers in Thanjavur may learn to grow paddy with less water or partially shift to other crops. As to how much should be allocated to

each State, each must have presented and documented its claims to Cauvery waters before the Tribunal, and also commented on profligate or improper use by the other. The Tribunal can be trusted to take all these arguments into account in its final Order.

(9) *Tamil Nadu has groundwater; it also has the benefit of the north-east monsoon*

What this means is that Tamil Nadu has other sources of water besides the Cauvery. That point must have been made by Karnataka before the Tribunal, and Tamil Nadu must have answered it. Without going into the arguments of the two States let us note that it would not be right to say to Tamil Nadu: 'You have groundwater and the north-east monsoon, so leave Cauvery alone.' The Cauvery is an inter-State river and all the States in the basin are entitled to make reasonable use of its waters. The Tribunal will surely take the totality of circumstances into account in making its allocations.

My final plea

At the conclusion of the meeting I made the following plea, which I would now like to repeat to a larger public in both the States:

(i) Let us not undermine the work of 12 years by questioning the legality of the Tribunal proceedings or the soundness of adjudication. These arguments are entirely wrong, but they have the potential of causing confusion and delay.

(ii) The RBA has been a dead letter. By all means try and revive it for the future, but it cannot be invoked in the present case.

(iii) Let us not waste too much time on the Interim Order. That Order will soon be replaced by the Final Order of the Tribunal.

(iv) Even at this stage, explore the possibility of bringing about a negotiated agreement, which can be reported to the Tribunal and converted into an Award. Non-official initiatives in this regard will be very useful.

(v) If need be, present supplementary material to the Tribunal if that is possible at this stage.

(vi) The Final Order is bound to be a fair, objective, carefully considered and fully argued Order. Wait for it and consider it with an open mind. If there are any doubts or dissatisfactions with it, a further reference can be made to the Tribunal within three months, and a supplementary report asked for.

(vii) Mount a campaign to dispel the miasma of suspicion, anger and misunderstanding that clouds public opinion in both States. Educate the general public—and in particular the farmers in the Cauvery basin—on the facts, the issues involved and the applicable principles.

(viii) More than anything else, try and prevent the further deterioration of the relations between the peoples of the two States. Build bridges, promote goodwill and understanding. Avoid the tragedy of a deep divide.

Appendix III

Some Notes on a Partial Study of Local Water-Harvesting and Watershed Development Initiatives

Note: During the period November 2000–October 2001, I was engaged (at Winrock International India, with a grant from the Ford Foundation) in a study of local, community-based water-harvesting and watershed development initiatives. As the first part of that two-year study, a preliminary round of field-visits was undertaken with a view to obtaining a better understanding of the nature of such initiatives and the operative factors and forces, as the preparation for a deeper and more intensive study in the second part. For personal reasons (mainly health-related) I was obliged to leave the project in October 2001. The project is continuing at Winrock and will in due course result in a set of findings and perhaps a book. Being no longer associated with the study, I do not know the direction and thrust of the second-phase of work and cannot guess what the findings will be. I await the outcome with great interest. However, the 'preliminary' round of field-visits in the first half of 2001 did provide some insights and raised some issues for consideration. In August–September 2001 I wrote a 'Discussion Paper' with the assistance of Radhika Gupta who was working with me on the project. That paper was discussed with a select group of experts at a meeting organized by Winrock on 27 September 2001, and the discussions at that meeting must have been of considerable assistance to Winrock in designing the second-phase study. As I found the preliminary round very interesting, I thought I should share my observations with the readers of this book. Extracts from the discussion paper of September 2001 are reproduced below.

Extracts from the Discussion Paper Prepared for Winrock International India in September 2001

Extract 1: Questions Formulated

What is the potential and significance of these activities as contributors of water? Are they bound to remain small, minor and supplementary, or are they in fact capable of becoming significant, even substantial, components of national water policy and planning? That split itself into three groups of sub-questions: (i) What has been the extent of increase in local water availability through these means in different places, and what have been the economic and social consequences of such increases? Do they have any offsetting negative consequences? (ii) What have been the operative factors? How do such initiatives start and how do they proceed to success or failure? What forces and factors help or hinder them? How sustainable are such developments over time? (iii) How replicable are such approaches in different physical areas and different economic, social, cultural, political and institutional environments and circumstances?

Extract 2: Places Visited

(1) Region, State, Place	(2) Climatic Conditions	(3) Agency (Governmental)	(4) Agency (NGO)	(5) Leadership
North: Haryana (Sukhomajri, Dhamala, Raelmajri)	Arid to semi-arid	State Forest Dept., Central Soil and Water Research and Training Institute		The late P.R. Mishra
North: Rajasthan, Alwar district (several villages)	Arid (less than 400 mm of rainfall)		Tarun Bharat Sangh	Rajendra Singh

Central: Madhya Pradesh (Sagar district, Jhabua district)	Arid	Government of Madhya Pradesh (Rajiv Gandhi Watershed Mission, *Paani Roko Abhiyan*)	Chief Minister, individual officials in the Secretariat, individual Collectors
Central: Madhya Pradesh (Bagli district)	Arid	*Samaj Pragati Sahayog* (RGWM, CAPART funds)	
Central: Madhya Pradesh (Indore)			
West: Maharashtra (Ralegan Siddhi, Ahmadnagar district)	Arid to semi-arid	Hind Swaraj Trust (with co-operation from the Govt. and funds from Plan programmes)	Anna Hazare
West: Maharashtra (Hivre Bazaar, Ahmadnagar district)	Arid to semi-arid	*Panchayat* (*Adarsh Gaon Yojana*)	Sarpanch Popatrao Pawar
West: Maharashtra	Arid to semi-arid	WOTR, Ahmadnagar	Father Bacher in

(areas covered by Indo–German Foundation programme)		the early stages	
N.B.: No field visits; discussions with Father Crispino Lobo			
West: Maharashta, Pune (No field visit, discussions at Pune)	Arid to semi-arid	*Paani Panchayat*	Vilasrao Salunke
South: Andhra Pradesh (Kadiri watershed, Anantapur district)	Arid	MYRADA, Bangalore	
South: Tamil Nadu (Kottampatti, Madurai district, Theni district)	Low to Medium rainfall (700– 800 mm.)	Dhan Foundation, Madurai	
Also discussions at Madurai with MYRADA			

South: Tamil Nadu, discussions at Chennai with MIDS (A. Vaidyanathan, S. Janakarajan), Madras School of Economics (Paul Appasamy), Anna University (Dr Karmegham) and A. Mohanakrishnan, retired Engineer-in-Chief of Tamil Nadu			
East: West Bengal (Purulia)	Heavy rainfall	West Bengal Government, (Jhalda Block Development Office)	
East: West Bengal (Tatanagar, Birbhum) Also discussions at Calcutta with LWS			Lutheran World Service

N.B.: The separation between 'Government' and 'NGOs' in the table above needs to be qualified. Whichever of the two be the initiating agency, it generally finds it useful to enlist the support or cooperation of the other.

Extract 3: What We Saw and Heard

(1) *Water-Related Points*

    (i) In all the areas that we visited, water harvesting or watershed development had been driven by a scarcity of water: either a constant state of water scarcity as in arid or drought-prone areas, or seasonal scarcity or unreliability even in areas where rainfall is moderate to heavy. (Periodical water scarcity can be experienced even in areas of very heavy rainfall, as for instance in Cherrapunji, known as the wettest place on Earth with an annual rainfall of over 11,000 millimetres (mm), but suffering from a severe seasonal water shortage because of the rapid run-off. The Centre for Science and Environment (CSE) has suggested that water harvesting could help even Cherrapunji.)

    (ii) A modicum of precipitation might seem a pre-condition for successful rainwater harvesting, but evidently the limit is fairly low. Watershed development and water harvesting have been successfully undertaken even in areas where the average rainfall is lower than 400 mm a year (e.g., Alwar district in Rajasthan, Jhabua district in Madhya Pradesh). It would appear that (purely from the point of view of a threshold of rainfall, and ignoring other relevant aspects) this activity is feasible in other parts of the country where the rainfall is of the order of around 300 mm or higher. The actual extent of water harvested will of course vary according to the soil conditions and topography of each area.

    (iii) In some cases, external sources of water are complementing water made available through water harvesting and watershed development. In Ralegan Siddhi in Maharashtra for instance, the water drawn from the Kukdi canal must be taken into account.

    (iv) In some areas, traditions of water harvesting or tank irrigation systems have gone into relative decline and people have resorted to other measures such as the increased mining of groundwater through tubewells (e.g., in Tamil Nadu). In such cases, it may not be easy to revert to the old systems, particularly where (as in the areas where the Dhan Foundation is working) there have been not merely the neglect and decline of tank systems, but also changing demographic and land-use patterns. However, the effort needs to be made, and is being made.

(v) While it is clear that more water has become available for use in the places visited, it is difficult to quantify this. Data on the availability of water before and after the intervention is generally not available.

(2) *Benefits*

(i) What was strikingly noticeable in all the cases was an increase in agricultural production and an enhancement in livelihoods (including milk production and holdings of livestock). Even in arid areas, water-harvesting has made possible cultivation where there was none before, if only ɔ meet subsistence needs. In some areas (the Tarun Bharat Sangh (TBS) area, for instance) greenery, agricultural activity and the continuing presence of water in the wells, though at low levels, could be observed even after three successive droughts.

(ii) The incidence of benefits from water harvesting varies across groups, such as the landowners, the landless (usually working as wage labour), and the cattle owners (gra. -). For the last two categories the benefits of watershed development are not always direct and assured, but they are real and can be significant. For the landowners, the main use of the increased availability of water is for irrigation. For wage labourers, the availability of work on the construction of structures within their area or village, or growth in farm-labour opportunities arising from increased agricultural activity, reduces the need to travel long distances in search of work. For those maintaining cattle or goats, rainwater harvesting in communal structures and increase in biomass from catchment-area treatment (to protect those structures from siltation) provide drinking water and fodder for livestock, leading to the increased production of milk for sale and growth in dairy development. This was clearly evident in Ralegan Siddhi and Hiwre Bazaar. Moreover, as watershed development (especially the 'ridge-to-valley' approach) often necessitates a ban on grazing in the catchment, an inducement or compensation has been provided in some cases in the form of permission for the sale of (saleable varieties of) grass, which increases in availability because of that protective measure. For instance, in Sukhomajri, the community has in the past benefited from the sale of bhabbar grass, the production of which was made possible through the work of watershed development (though the saleability of that grass has declined in recent years). Similar benefits could also be observed in the TBS areas in Rajasthan.

(iii) To the extent that water harvesting through various structures (*johads*, earthen check dams, tanks in south India, etc.,) leads to the recharge of groundwater aquifers (which can be seen in the rising of water levels in wells, or the availability of water in wells even in times of drought), there is a direct correlation between water harvesting/watershed development and the availability of drinking water. (The availability of drinking water from the municipality, and the sources of that water, will of course need to be taken into account.)

(iv) In the course of water harvesting and watershed development efforts many issues of equity and justice arise. With the exception of Sukhomajri and Dhamala (in Haryana), where the landless have equal rights to water, and are free to sell their share of it, and the '*Paani Panchayat*' in Pune, which has adopted the principle of delinking land and water rights, it seemed clear that the landed stood to benefit most from local water augmentation. The landless, as mentioned earlier, do benefit, but indirectly, from the improvement of common property resources.

(v) Distortions and inequities in the incidence of the benefits of increased water availability arising from water harvesting and watershed development can result from the freedom of landowners to exploit the groundwater in their lands by virtue of the easement rights to such water that vest in them under the law as it stands at present. (In Maharashtra WOTR has been trying to get tubewells banned, but has met with little success in the absence of any legislation.)

(It could be argued that issues of water distribution and equity arise only after water has been generated, and that to judge any water-related intervention *ab initio* on the grounds of equity and justice would be premature. However, in the post-augmentation stage one must definitely confront this issue.)

(3) *Leadership and Approaches*

The places visited could be broadly grouped under three heads: (i) individual initiatives, (ii) governmental initiatives, and (iii) the programmes of NGOs. Inspirational leadership undoubtedly plays an important role in the first category. Anna Hazare's name is ineluctably linked with the story of Ralegan Siddhi, and the name of Rajendra Singh with the work of the TBS. It is clear that these two success stories can be attributed to the exemplary motivation and zeal that these leaders brought to social/community mobilization in their respective areas. In purely government-fostered initiatives such as those in Madhya Pradesh, community mobilization is more institutionalized and less dependent on the presence of an individual with leadership qualities in a village, though over time some sort of leadership seems to emerge naturally within any community. However, even in such cases, the presence of an enthusiastic, committed, energetic Collector or Commissioner[61] makes a vital difference. We came across an instance of this in Madhya Pradesh. Similarly, the evident success of the Hivre Bazaar story in Maharashtra is entirely due to the commitment, enthusiam and energy of the Sarpanch Popatrao Pawar. In Haryana, the late P.R. Mishra, formerly director of the Central Soil and Water Conservation Research and Training Institute, is widely acknowledged as having been the inspirational leader behind the Sukhomajri success story. (Even today, the senior officer concerned in that Institute and the present Principal Chief Conservator of Forests of the State are dedicated officials.) However, we also have other cases where no such 'prime mover' can be identified. These are interventions of the third kind: that of non-governmental organizations. In the areas where the Dhan Foundation has worked, for instance, success is not dependent on one visionary local leader or an exceptional government official. Leadership seems to emerge gradually within the Tank Farmers' Associations.

It may be useful to note here the differences in approach among the various instances mentioned above. Anna Hazare was inspired by Gandhi and Vivekananda and his earnest desire was to bring about a moral and social transformation; water was an important entry point and continues to engage his attention, but that interest is subsumed under his larger concerns. Sarpanch Popatrao Pawar of Hiwre Bazaar was doubtless inspired by Anna Hazare, but his approach seems to be more managerial (an ability to make the official machinery work and

---

61 This personal motivation may be an exceptional individual phenomenon, or it may be inspired by and reflect the visionary approach, zeal and high commitment at the highest political level in the State.

to make people listen to him) than moral. However, as an elected leader of his village, his own personal motivation has charged the work of watershed development, and he too is driven by a pride in his village and a desire to enhance not merely its economic status but also its reputation and social standing. On the other hand, the leadership provided by Father Crispino Lobo in the areas where WOTR has been working seems more impersonal in nature, guided by the objectives of the organization. The community's relationship with him would probably be very different from that of Ralegan Siddhi with Anna Hazare. He also does not feel that it is necessary to build up and depend on strong local leadership. He believes in providing help to those villages which wish to help themselves. However, it would be wrong to minimize his deep commitment to the cause and his concern for human need, equity, and good resource management.

(4) 'Replication' or Spread Effect

An effort on the part of the Maharashtra Government to replicate the Ralegan Siddhi model in 300 other villages through a committee under Anna Hazare's chairmanship appears to have failed. On the other hand, Hivre Bazaar is in a sense a replication of Ralegan Siddhi, because Sarpanch Pawar was motivated and inspired by that example. In Rajasthan, it has been stated that the movement started by the TBS now extends to 700 villages. That figure has been questioned by critics who say that it is an exaggeration, but there is no doubt that the TBS's efforts in Alwar district have had some spread effect. NGOs such as Paani Panchayat and WOTR in Maharashtra, Pradan in the North and East, and MYRADA and Dhan Foundation in the South, have also stretched their coverage slowly and to a modest extent. In Madhya Pradesh the Government is steadily expanding the coverage of its watershed development activities from a limited number at the start to a few thousands now, and the ambitious plan now is to bring the entire State within the ambit of the *Paani Roko Abhiyan*. Prima facie, it would appear that while methods, techniques, institutional arrangements, and so on, will vary from place to place, the broad approach of the local harvesting and conservation of water has been and can be replicated.

(5) Difficulties, Decline?

There have been reports of a 'decline' in Sukhomajri, but this is disputed by the PCCF who contends that even with the passage of time the motivation of the community has not significantly declined. He agrees, however, that some difficulties have emerged. The first is the fact that one important source of income and motivating factor, namely the sale of bhabbar grass collected from the protected (socially fenced) catchment area has all but disappeared because the principal buyers, namely the paper mills in the neighbourhood, have largely switched to an alternative material, i.e., eucalyptus. This is a technological development that was not foreseen. An alternative market for the grass (or an alternative income-generating activity) has not yet been identified. This is a serious and as yet unresolved difficulty. Secondly, out of the large number of check-dams built in the 1980s several have been affected by siltation. This was not an unforeseen problem, but the problem has not been dealt with satisfactorily as yet. Thirdly, while afforestation in the catchment initially improved the availability of water by retarding the run-off, it eventually resulted in reduced flows into the check-dams. The answer is said to lie in a balancing of trees and grass in the catchment, as trees retard the flow of both water and top-soil, whereas grass retards the latter more than the former. Further, there is a longstanding conflict of interests between Sukhomajri and Dhamala. Undoubtedly, these two 'successful' villages are beset with some difficulties, but none of them seems insurmountable. In Ralegan Siddhi the very success of the effort may be creating problems. The younger generation have grown up in a situation of prosperity, which they tend to take for granted. They cannot recall the bleak and desperate situation faced by the older generation, and cannot be expected to have the same zeal to work hard for the common good or to accept sacrifices in the process. They are also likely to be susceptible to the blandishments of urban lifestyles. The future of Ralegan Siddhi seems uncertainly poised.

(6) *Tradition and Modernity*

In many cases, rainwater harvesting initiatives involve the revival of age-old traditions, but this is not true in all the cases. While the repair of *johads* in Rajasthan and the rehabilitation of tank systems in south India may be instances of the revival of old traditions, the concept of water harvesting appeared to be fairly new in some parts of Madhya Pradesh, where drought propelled the need for it. In Haryana (Sukhomajri), the primary objective to start with was the stopping of the siltation of the Sukhna lake through the treatment of the catchment area; water harvesting was an incidental development and an incentive for the people in the region to conserve the catchment. Watershed treatment itself, especially the holistic 'ridge-to-valley' approach, is a modern, technically worked out idea. Within that framework, old traditions of water harvesting, if extant, could be incorporated. Water harvesting brings to the fore the hydro-geological specificity of each area. The combination of tradition with innovation is important if old water harvesting systems are to be successfully revived so as to be responsive to present day needs. However, tradition can be an important tool for invoking collective memory for social mobilization. It can also be an important assertion of the functioning of customary rules, as opposed to state-imposed norms.

(7) *Institutional Development*
The work of water harvesting and watershed development has fostered institutional development within communities across all the cases we visited. Similarities can be discerned amongst all the cases in the formation of some sort of association/committee/samiti, to govern the actual work as well as the distribution and management of water, as also the evolution of micro-credit and self-help groups. However, the degree of institutional development (in terms of the financial strength of the institutions, their ability to interact with state agencies and donor organizations effectively and independently, and the overall structure and systems evolved) varies from case to case. For instance, the samitis (also known as the 'gram sabhas') in Alwar (TBS area) still appear to be fledgelings in comparison with the degree of institutional development of the Tank Farmers' Associations in South India, which have federated into a 'district federation', as a stronger forum for dealing with external agencies.

(8) *Role of Women*
In some of our visits we noted that women were present at the meetings of watershed committees and gram sabhas, and that some of them participated in the deliberations. However, it was difficult to judge to what extent this was an indication of real and effective participation by women in the processes of decision-making. This will need to be probed further.

(9) *Source of Funds*
There are variations in regard to the funding of such activities. The funds come from Plan programmes, World Bank loans, Indian

foundations, foreign donor agencies, etc. These variations by themselves did not seem to cause significant differences, but this again is a matter that needs to be gone into further.

(10) *Coordination*

The lack of inter-sectoral coordination hinders the effective implementation of watershed-development programmes. For instance, the Forest Conservation Act, if literally and rigidly enforced, could hinder an integrated ridge-to-valley treatment approach, if the ridge areas fall within the purview of the Forest Department. In governmental programmes, tasks which form part of an integrated whole are undertaken separately by different Departments—such as the Soil Conservation Department, the Department of Agriculture and so on—with community-mobilization often being treated as a separate if parallel process. This creates confusion at the ground level, and calls for a review of implementation strategies.

(In general, the success of community initiatives in the area of water harvesting and watershed development is often helped or hindered by policies and laws that have no direct connection with them but nevertheless impinge on them ... such as agricultural subsidies which may promote cropping patterns that are more water-intensive, subsidies on diesel which may result in the over-extraction of groundwater, etc.)

(11) *Watershed Committees and Panchayats*

It is noteworthy that watershed development work in many of the places visited is being undertaken not through the local panchayats but through the constitution of independent water users' associations/tank farmers' federations/village watershed-development committees. (This has been understood as being 'administrative devolution' in terms of the 1994 'Common Guidelines', as opposed to statutory or constitutional decentralization.) The politicization of *panchayats*, and the fact that *panchayats* do not constitute ecological units (with the result that villages falling under a certain *panchayat* may not correspond to the boundaries of a watershed), are the two principal reasons cited for this situation. The relationship between *panchayats* and the committees/associations mentioned above varies from case to case, being more difficult in some places than in others. The sharing of usufructs between tank farmers' associations and the *panchayats* has been problematic in south India. On the other hand, while villagers may not want to undertake watershed work through the *panchayat*, as it is seen as being overly politicized, *panchayat* members are generally co-opted ex-officio into the committees/samitis/associations formed. (It is perhaps ironic that over-politicization becomes a hindrance to greater *panchayat* involvement with watershed and water harvesting work, as *panchayats* can be effective vehicles for political will at a larger level.) An illustration of the right kind of link could perhaps be found in the work of PRADAN in West Bengal. It has been successfully trying to foster increasing *panchayat* involvement. Similarly in Madhya Pradesh, the figure of Chief Minister Digvijay Singh (and therefore the State) looms large in any discussion of development work. Other instances of positive panchayat involvement have been found in Hiwre Bazaar and Ralegan Siddhi. The Dhan Foundation too is exploring the idea of advocating the formation of sub-committees on tanks under the panchayat in the future, in order to enable tank farmers to gain usufruct rights.

(Most water users' associations/village watershed-development-committees/tank farmers' associations have been registered under an Act of the State—often the Indian Societies Registration Act, but sometimes other Acts, and in Andhra Pradesh a special Act—in order to give them a legal status. However, this confers only limited powers on them. Panchayats, as the third tier in the federal structure brought in by the 73rd Constitutional Amendment, have a statutory role to play, depending on the extent of devolution actually carried through, and have legal rights over usufructs. They are in fact 'state' at that level. They have the power to constitute statutory bodies with the responsibility of handling government funds, collecting revenues and delegating responsibilities. It will be necessary to examine what functions and powers have in fact been devolved to *panchayats* in each State, particularly in relation to natural resource-management. It is important to develop a good working relationship between the watershed committees/*samitis*/associations and the *panchayats*. At the same time, from the point of view of equity and local power relations, there is the danger that the processes of *panchayati raj* may help the local power elite to consolidate their power further to the detriment of weaker groups or sections within a village or a community.)

(12) *Community Initiatives and the State*

(i) In practical terms, funds from donor agencies available for development still constitute only a fraction of the total funds needed, ensuring that the Government, both at the Central and State level, remains significant. Even though the savings funds of watershed committees/associations, etc., are important sources for communities to draw from for repair and maintenance work, there still seems to be a need for outside help when it comes to major repairs. In the event of the withdrawal of a development agency or the closure of a specific project, communities/villages are left only with the local government as a source of support. We found that in Mayurbhanj (Orissa), where the Lutheran World Service (LWS) had implemented the construction of water-harvesting structures under an integrated rural development programme, the grain banks and savings groups formed for looking after maintenance work were still functional five years after the withdrawal of the LWS, but the people were dependent on the Block Development Office for funds if any major repair work had to be undertaken. Under the 'Rajiv Gandhi Watershed Mission' (RGWM) in Madhya Pradesh for instance, the funds being diverted to the 'Village Watershed Committees' are nothing but the 'greening' of the existing Department for Rural Development funds for rural development.

(ii) The emergence of the 'Arwari Parliament' in Rajasthan in the context of a dispute with the State government over the licensing of fishing raises important issues concerning the relationship between state and civil society that will need further discussion.

(This issue, which appeared to have become quiescent, has once again come to the fore with the recent questioning by the Irrigation Department of the legality of a water-harvesting structure built by the villagers. This confrontation between civil society and the state seems ominous. The principle of 'people's participation' is enshrined in the 'Common Guidelines' of 1994, which was an enlightened document considerably ahead of its time. Anna Hazare received significant support from the agencies of the Maharashtra Government. The Madhya Pradesh Government is actively enlisting the cooperation of the people and NGOs in its RGWM and *Pani Roko Abhiyan* programmes. At this juncture, the assertion of the Rajasthan Irrigation Minister that every drop of rain belongs to the state and that the water-harvesting activities of the people are illegal seems very retrograde and is fraught with serious consequences. It is to be hoped that this setback to the movement will be quickly removed.)

Select References and Bibliography

Abbas, B.M. (1984), *The Ganges Waters Dispute*, University Press Ltd., Dhaka.

Adhikary K.B., Q.K. Ahmad, S.K. Malla, B.B. Pradhan, Khalilur Rahman, R. Rangachari, K.B. Sajjadur Rasheed, B.G. Verghese, eds. (2000), *Cooperation on the Eastern Himalayan Rivers: Opportunities and Challenges*, Konark Publishers, New Delhi, under the auspices of Bangladesh Unnayan Parishad, Dhaka, Centre for Policy Research, New Delhi, and Institute for Integrated Development Studies, Kathmandu.

Ahmad Q.K., Asit K. Biswas, R. Rangachari, M.M. Sainju, eds. (2001), *Ganga-Brahmaputra-Meghna Region: A Framework for Sustainable Development*, The University Press Ltd., Dhaka.

Asian Development Bank (1998), '*The Bank's Policy on Water: Working Paper*, August 1998.

Barh Mukti Abhiyan (1997), *Proceedings of the Second Delegates Conference*, Patna.

Baxi, Uppendra (2002), *The Future of Human Rights*, Oxford University Press, New Delhi.

Benda-Beckmann, Franz von (2001), 'Water, Human Rights and Pluralism', first draft, presented at the Conference on Water Resources, Human Rights and Governance organized by the Nepal Water Conservation Foundation at Kathmandu from 26 February 2001 to 2 March 2001 (unpublished).

Biswas, Jellali and Stout (1993), *Water for Sustainable Development in the 21st Century*, Oxford University Press, Delhi.

Bruns, Bryan Randolph, and Ruth S. Meinzen-Dick, eds. (2000), *Negotiating Water Rights*, IFPRI, Vistaar Publications, New Delhi.

Centre for Science and Environment, New Delhi:
State of India's Environment: The First Citizens' Report, (1982).
State of India's Environment: The Second Citizens' Report, 1985.
State of India's Environment: The Third Citizens' Report: Floods, Floodplains and Environmental Myths, (1991).
Dying Wisdom: Rise, Fall and Potential of India's Traditional Water Harvesting Systems, (1997).
Drought? Try Capturing the Rain, Anil Agarwal, (2000).
Making Water Everybody's Business: Practice and Policy of Water Harvesting, Eds. Anil Agarwal, Sunita Narain and Indira Khurana, March 2001.

Chauhan, B.R. (1992), *Settlement of International and Inter-State Water Disputes in India*, under the auspices of the Indian Law Institute, N.M. Tripathi Pvt. Ltd., Bombay, 1992.

Chitale, M.A. (2000), *I Predict A Blue Revolution*, Bharatiya Vidya Bhavan, Pune.

Chopra, Kanchan and Biswanath Goldar (2000), 'Sustainable Development Framework for India: The Case of Water Resources'–Final Report, Institute of Economic Growth, Delhi, for the UN University, Tokyo, October 2000.

Chopra, Kanchan, Gopal K. Kadekodi and M.N. Murty (1998), 'Sukhomajri and Dhamala Watersheds in Haryana: A Participatory Approach to Management' May 1998, Institute of Economic Growth, Delhi.

Crow, Ben with Alan Lindquist and David Wilson (1995), *Sharing the Ganges: The Politics and Technology of River Development*, Sage Publications, New Delhi.

Crow, Ben (1997), 'Bridge over Troubled Waters? Conflict and Cooperation over the Waters of South Asia' in *Regional Cooperation in South Asia: Prospects and Problems*', ed. Sony Devabhaktuni, Occasional Paper No. 32, February 1997, The Henry L. Stimson Center, Washington D.C., USA.

Datye, K.R. (1994), Narmada-Outline of an Alternative, Notes for presentation to the Five Member Group on Various Issues Relating to the Sardar Sarovar Project, Report, Vol. II (Appendices).

Dhawan, B.D. (1990), ed., *Big Dams, Claims, Counter Claims*, Commonwealth Publishers, New Delhi.

(1993), *Indian Water Resource Development for Irrigation: Issues, Critiques and Reviews*.

Down to Earth (Journal of the Centre for Science and Environment), 15 November 2002, article entitled 'Overflowing Passions in Cauvery'.

Dreze J., Meera Samson and Satyajit Singh, eds. (1997), *The Dam and the Nation: Displacement and Resettlement in the Narmada Valley*, Oxford University Press, Delhi.

Dubash, Navroz (2002), *Tubewell Capitalism: Groundwater Development and Agrarian Change in Gujarat*, Oxford University Press, New Delhi.

Farrington John, Cathryn Turton, and A.J. James, eds. (1999), *Participatory Watershed Development: Challenges for the 21st Century*, Oxford University Press.

Gaur, Vinod K., ed., (1993), *Earthquake Hazard and Large Dams in the Himalayas*, INTACH, New Delhi.

Gleick, Peter:
   (1996), 'Basic Water Requirments for Human Activities: Meeting Basic Needs', *Water International* 21 (1996), International Water Resources Association.
   (1998), 'The Human Right to Water', *Water Policy I* (1998), Elsevier Science Ltd. 1999.

Goldsmith E. and N. Hildyard (1984), *The Social and Environmental Impacts of Large Dams*, Vol. 1 & 2, Wadebridge Ecological Centre, Cornwall.

Government of India:

'National Water Policy', Ministry of Water Resources, 1987.

'Report of the Commission on Centre-State Relations (Sarkaria Commission)', Government of India Press, Nasik, 1987–88.

'Organizational and Procedural Change Requirements in the Irrigation Sector', Central Water Commission, 1992.

'Reassessment of Water Resources Potential of India', Central Water Commission, 1993.

'Guidelines for Watershed Development', Ministry of Rural Areas and Employment, 1994.

'Report of the Five Member Group on Various Issues Relating to the Sardar Sarovar Project', Ministry of Water Resources, 1994.

'Further Report of the FMG on Certain Issues Relating to the Sardar Sarovar Project', 1995 (unpublished).

'Report of the Expert Committee on the Environmental and Rehabilitation Aspects of the Tehri Hydro-Electric Project', Ministry of Power, New Delhi, 1997 (unpublished).

'Report of the Working Group on Participatory Irrigation Management for the Ninth Plan', Ministry of Water Resources, New Delhi, 1997.

'Report of the Working Group on Perspective of Water Requirements' (National Commission for Integrated Water Resources Development Plan), Government of India, Ministry of Water Resources, New Delhi, September 1999.

'Integrated Water Resources Development—A Plan for Action: the Report of the National Commission on Integrated Water Resources Development Plan', Ministry of Water Resources, September 1999.

'Reply to the World Commission on Dams', letter No.2/WCD/2001/DT (PR) Vol. III dated 1–2–2001 addressed to the Secretary General of the WCD (see WCD's Website www.dams.org).

'Comments on the Internet on WCD's Report', (see http://genepi.louis-jean.com/cigb/Inde.htm).

'National Water Policy 2002', Ministry of Water Resources, New Delhi, 2002.

Government of India, Tribunals' Awards:

Krishna Waters Dispute Tribunal, May 1976.*

Narmada Waters Dispute Tribunal, December 1979.

Godavari Waters Dispute Tribunal, July 1980.*

(*Date of notification of final awards in the Government of India Gazette).

Ravi–Beas Waters Dispute Tribunal–Report 1986 (not yet notified; further reference made to the tribunal).

Cauvery Waters Dispute Tribunal–Interim Order 1991 notified (proceedings in progress).

Government of India, Treaties:
The Indus Waters Treaty 1960.
The Mahakali Treaty 1996.
The Ganges Water Sharing Treaty 1996.

Guhan, S. (1993), *The Cauvery River Dispute: Towards Conciliation*, Frontline Publications, Madras.

Gulhati, N.D. (1973), *Indus Waters Treaty,* Allied Publishers Pvt. Ltd., New Delhi.

Gyawali, Dipak and Ajaya Dixit (2001), 'How Not To Do A South Asian Treaty', *Himal, South Asian Magazine*, Kathmandu, April 2001.

Gyawali, Dipak (2001), *Water in Nepal*, Himal Books, Kathmandu.

Hanumantha Rao, C.H. (2000), 'Watershed Development in India: Recent Experiences and Emerging Issues', *Economic and Political Weekly*, Bombay, 4 November 2000.

Himal, South Asian Magazine, Kathmandu, Nepal: Fantastic Dams (special issue), Vol. 11, Number 3, 1998.

Indian National Academy of Engineering (1990), *Water Management: Perspectives, Problems and Policy Issues*, New Delhi, 1990.

Indian Water Resources Society:
'1997: River Basin Management: Issues and Options', New Delhi.
'1998: Five Decades of Water Resources Development in India', New Delhi.

International Conference on Water and Environment, Dublin Statement, Dublin 1992.

International Law Association (1966), 'Helsinki Rules on the Uses of the Waters of International Rivers', International Law Association, London.

Iyer, Ramaswamy R.
See Appendix I.

Jain, L.C. (2001), 'Dams Versus Drinking Water: Exploring the Narmada Judgement', Parisar, Pune, 2001.

Jayal, N.D. (1993), *Ecology and Human Rights*, INTACH, 1993.

Lahiri Dutt, Kuntala:
'Imagining Rivers', *Economic and Political Weekly*, Bombay, 1 July 2000.
'Rebirth of a River', *Economic and Political Weekly, Bombay*, 14 April 2001.

McCully, Patrick (1998), *Silenced Rivers: The Ecology and Politics of Large Dams*, Orient Longman (India).

Menon, M.S., 'WCD Report: A Framework for Underdevelopment?' *The Hindu*, 14 August, 2001.

Mishra, Dinesh Kumar (1990), *Barh se trast–Sinchai se past: Uttar Biharki Vyatha Katha'* (Hindi), Samata Publications, Patna.

Mishra, Dinesh Kumar, 'Living with Floods: People's Perspective', Economic and Political Weekly, 21 July 2001.

Moench Marcus, Elisabeth Caspari and Ajaya Dixit, eds. (1999), *Rethinking the Mosaic: Investigations into Local Water Management*, Nepal Water Conservation Foundation, Kathmandu, and Institute for Social and Environmental Transition, Boulder, Colorado, USA, December 1999.

National Commission to Review the Working of the Constitution (2002), *Report*, Universal Law Publishing Co. Pvt. Ltd., New Delhi, 2002.

Navalawala, B.N., 'World Commission on Dams: Biased?', *Economic and Political Weekly*, 24 March 2001.

OIKOS and IIRR (2000), *Social and Institutional Issues in Watershed Management in India*, OIKOS India and International Institute of Rural Reconstruction, Y.C.James Yen Center, Silang, Cavite, The Philippines.

Pangare, Vasudha and Ganesh (1992), *'From Poverty to Plenty: The Story of Ralegan Siddhi'*, INTACH, New Delhi, 1992.

Parasuraman, S., 'WCD Report: Response to a Misreading' *The Hindu*, 11 September, 2001.

Petrella, Ricardo, *The Water Manifesto*, Zed Books, London, and Books For Change, Bangalore, 2001.

Planning Commission:
'Report of the Committee on the Pricing of Irrigation Water', New Delhi, 1992.
'Report of the Ninth Plan Working Group on Major/Medium Irrigation', New Delhi.
'Internal Papers on "Externally Aided Projects" and "Rural Water Supply"'.
'Approach Paper to the Tenth Five Year Plan (2002–2007)', September 2001.

Raina, Vinod, Presentation at WISCOMP Symposium on 'Conflict-Resolution: Trends and Prospects', New Delhi, 3–7 October 2001.

Rangachari, R., 'Some Disturbing Questions', *Seminar*, 478, June 1999.

Rangachari R., Nirmal Sengupta, Ramaswamy R. Iyer, Pranab Banerji, Shekhar Singh (2000), 'Large Dams: India's Experience, A Report to the World Commission on Dams'.

Rao, K.L. (1975), *India's Water Wealth: Its Assessment, Uses and Projections*, Orient Longman, New Delhi.

Reddy, Amulya K.N. and Girish, Sant, 'Electrical Part of the Sardar Sarovar Project', Submission to the Five Member Group on the Sardar Sarovar Project, (Report of the FMG,Vol. II, Appendices, 1994).

Roy, Arundhati (1999), *The Greater Common Good*, India Book Distributors.

Seminar (a New Delhi journal) No. 478, June 1999, Issue on 'Floods: A Symposium on Flood Control and Management'.

Sengupta, Nirmal 1985, 'Irrigation: Traditional *versus* Modern', *Economic and Political Weekly*, Vol. 20, No. 45, 46, 47, 1985.

1991, *Managing Common Property: Irrigation in India and the Philippines*, Indo–Dutch Studies on Development Alternatives, Sage Publications, New Delhi.

1993, *User Friendly Irrigation Designs*, Sage Publications, New Delhi and London.

2001, 'Report of the World Commission on Dams: Biased if Misread', *Economic and Political Weekly*, Bombay, 12 May 2001.

See also R. Rangachari, Nirmal Sengupta *et al.* above.

Shah, Mihir and others (1998), *India's Dry Lands*, Oxford University Press, Delhi.

Shah, Mihir (2002), 'Water Policy Blues', *The Hindu*, 7 June 2002.

Shah, R.B. (1994), 'Inter-State Water Disputes: A Historical Review', *International Journal of Water Resources Development*, Vol. 10, No. 2, Oxford, UK, 1994.

Shah, Tushar (1993), *Groundwater Markets and Irrigation Development*, Oxford University Press, Bombay.

Shankari, Uma and Esha Shah (1993), *Water Management Traditions in India*, PPST Foundation, Madras.

Sharma, Kalpana (2000), 'Guidelines for Future of Dams', *The Hindu*, 17 November 2000.

Shiva, Vandana:

(1998), 'Development, Ecology and Women' in *Debating the Earth: The Environmental Politics Reader*, eds. John S. Dryzek and David Schlosberg, Oxford University Press, 1998.

(2001), Presentations at: (i) International Conference on Globalization, Environment and People's Survival organized by the Research Foundation for Science, Technology and Ecology, New Delhi, 29 September 2001– 1 October 2001, and (ii) WISCOMP Symposium on 'Conflict Resolution: Trends and Prospects', New Delhi, 3–7 October 2001.

Water Wars: Privatization, Pollution and Profit, India Research Press, New Delhi, (2002).

Singh, Chattrapati:

(1991), *Water Rights and Principles of Water Resources Management*, Tripathi (for the Indian Law Institute), Bombay.

(1992), ed., *Water Law in India*, The Indian Law Institute, 1992.

Singh, Satyajit (1997), *Taming the Waters: The Political Economy of Large Dams in India*, Oxford University Press.

Thanh, N.C., and Asit K. Biswas (1990), *Environmentally Sound Water Management*, Oxford University Press, Delhi.

Thukral, E.G. (1992), *Big Dams, Displaced People: Rivers of Sorrow, Rivers of Change*, Sage Publications, New Delhi.

United Nations: UN Convention on the Non-Navigable Uses of International Water Courses, (passed by the UN General Assembly, 1997).

Vaidyanathan, A.:
1991, 'Integrated Watershed Development: Some Major Issues', Society for the Promotion of Wastelands Development, Foundation Day Lecture, May 1991.

1999, *Water Resources Management: Institutions and Irrigation Development*, Oxford University Press, New Delhi, 1999.

Valsalan, V.M. (1997), 'Inter-State Water Disputes in India: A New Approach' Central Board of Irrigation and Power, New Delhi, 1997.

Verghese, B.G.:
1990, *Waters of Hope: Himalaya–Ganga Development and Cooperation for a Billion People*, Oxford & IBH Publishing Co. Pvt. Ltd., for Centre for Policy Research, New Delhi, 1990.

1999, Second edition, with the sub-title *From Vision to Reality in Himalaya-Ganga Development Cooperation*.

1994, *Winning the Future*, Konark Publishers Pvt. Ltd. New Delhi.

Verghese, B.G. and Ramaswamy R. Iyer (1993), *Harnessing the Eastern Himalayan Rivers: Regional Cooperation in South Asia*, Konark Publishers Pvt. Ltd., for Centre for Policy Research, New Delhi.

Verghese, B.G., Ramaswamy R. Iyer, Q.K. Ahmad, S.K. Malla, and Pradhan, B.B., eds. (1994), *Converting Water into Wealth*, Konark Publishers Pvt. Ltd., New Delhi (also published simultaneously at Dhaka and Kathmandu).

Vohra, B.B. (1990), *Managing India's Water Resources*, INTACH, New Delhi.

WATER NEPAL, journal of water resources development, Kathmandu, Nepal, Vol. 6, No. 1, January–July 1998 (special issue on water resources development).

Water Vision 2025 papers:
Bangladesh Water Partnership, 'Bangladesh Water Vision', June 1999.

India Water Partnership, 'India Water Vision', July 1999 (Published by India Water Partnership and Institute for Human Development, New Delhi, 2000).

Jalasrot Bikas Sanstha, 'Nepal Water Vision', June 1999.

Pakistan Water Partnership, 'Pakistan Country Report–Vision for Water for the 21st Century', 15 June 1999.

'Sri Lanka Water Vision', prepared for the South Asia Regional Water Vision Workshop, 26–27 June1999.

World Bank:
'India: Irrigation Sector Review', Vol. 1 and 2, July 1991.

'Water Resources Management—A World Bank Policy Paper', 1993.

'India–Water Resources Management Sector Review', March 1998.

India Water Resources Management (set of six publications), The World Bank and Allied Publishers, in collaboration with GOI, Ministry of Water Resources, 1999.

'Round Table on Water Sector Strategy Review', New Delhi, 11–12 May 2000, WB's Website www.worldbank.org Topics and Sectors–Environment–Water Resources Management–Water Resources Management–Water Resources Strategy–South Asia).

'Water Resources Sector Strategy Review Draft', March 2002.

World Commission on Dams, Report, *Dams and Development: A New Framework for Decision-Making*, Earthscan Publications Ltd., London and Sterling, VA, November 2000.

World Bank and IUCN, 'Large Dams: Learning from the Past, Looking at the Future', July 1997.

World Humanity Action Trust (WHAT), 'Governance for a Sustainable Future, IV. Working with Water,' UK, 2000.

World Water Forum (The Hague) Papers:

World Water Vision Commission Report, 'A Water Secure World–Vision for Water, Life and Environment', World Water Council 2000.

Ministerial Declaration of The Hague, 'Water Security in the 21st Century', March 2000.

Names Index

Subject Index

About the Author

Ramaswamy R. Iyer is Honorary Research Professor at the Centre for Policy Research, New Delhi. In a civil service career spanning 34 years, he served in different capacities in various ministries and organizations of the Government of India before retiring in 1987 as Secretary, Ministry of Water Resources. He has also served as a consultant for various international bodies, including the World Bank and the World Commission on Dams, besides being the chairman or member of many committees, commissions and working groups set up by the Government of India, including being a member of the National Commission for Integrated Water Resources Development Plan (1996–99). Mr Iyer has published four books previously including *A Grammar of Public Enterprises*, *Harnessing the Eastern Himalayan Rivers* (co-edited) and *Converting Water into Wealth* (co-edited). He has also contributed chapters to edited volumes and written articles and essays on water resources, the environment, public enterprises, public administration, management and economic policy.